Pye Henry Chavasse

Advice to a Mother on the Management of her Children and on the

Treatment on the Moment of some of their More Pressing Illnesses

and Accidents

Pye Henry Chavasse

Advice to a Mother on the Management of her Children and on the Treatment on the Moment of some of their More Pressing Illnesses and Accidents

ISBN/EAN: 9783337764463

Printed in Europe, USA, Canada, Australia, Japan

Cover: Foto ©berggeist007 / pixelio.de

More available books at **www.hansebooks.com**

ADVICE TO A MOTHER

ON THE·

MANAGEMENT OF HER CHILDREN

AND ON THE

TREATMENT ON THE MOMENT

OF SOME OF THEIR MORE PRESSING ILLNESSES
AND ACCIDENTS.

BY

PYE HENRY CHAVASSE,

FELLOW OF THE ROYAL COLLEGE OF SURGEONS OF ENGLAND, FELLOW OF THE
OBSTETRICAL SOCIETY OF LONDON, FORMERLY PRESIDENT OF QUEEN'S
COLLEGE MEDICO-CHIRURGICAL SOCIETY, BIRMINGHAM.

"Lo, children and the fruit of the womb are an heritage and gift
that cometh of the Lord."

CANADIAN COPYRIGHT EDITION.

TORONTO:
WILLING & WILLIAMSON,
10 & 12 KING STREET EAST.
1880.

PREFACE.

THIS Book has been translated into French, into German, into Polish, and into Tamil (one of the languages of India); it has been extensively published in America; and is well-known wherever the English language is spoken.

The Twelfth Edition—consisting of twenty thousand copies—being exhausted in less than three years, the THIRTEENTH EDITION is now published.

One or two fresh questions have been asked and answered, and two or three new paragraphs have been added.

PYE HENRY CHAVASSE.

214, HAGLEY ROAD, EDGBASTON,
BIRMINGHAM, *June,* 1878.

CONTENTS.

PART I.—INFANCY.

PART II.—CHILDHOOD.

PART III.—BOYHOOD AND GIRLHOOD.

ADVICE TO A MOTHER.

Infant and suckling.—1. Samuel.
A rose with all its sweetest leaves yet folded.—Byron.
Man's breathing Miniature!—Coleridge.

PRELIMINARY CONVERSATION.

1. *I wish to consult you on many subjects appertaining to the management and the care of children : will you favour me with your advice and counsel ?*

I shall be happy to accede to your request, and to give you the fruits of my experience in the clearest manner I am able, and in the simplest language I can command— freed from all technicalities. I will endeavour to guide you in the management of the health of your offspring ;— I will describe to you the *symptoms* of the diseases of children ;—I will warn you of approaching danger, in order that you may promptly apply for medical assistance before disease has gained too firm a footing ;—I will give you the *treatment* on the·moment of some of their more pressing illnesses—when medical aid cannot at once be procured, and where delay may be death ;—I will instruct you, in case of accidents, on the *immediate* employment of remedies—where procrastination may be dangerous ;—I will tell you how a sick child should be nursed, and how a sick-room ought to be managed ;—I will use my best energy to banish injurious practices from the nursery ;—I will treat of the means to prevent

disease where it be possible ;—I will show you the way
to preserve the health of the healthy,—and how to
strengthen the delicate ;—and will strive to make a
medical man's task more agreeable to himself,—and more
beneficial to his patient,—by dispelling errors and prejudices, and by proving the importance of your *strictly*
adhering to his rules.   If I can accomplish any of these
objects, I shall be amply repaid by the pleasing satisfac-
tion that I have been of some little service to the rising
generation.

2. *Then you consider it important that I should be
made acquainted with, and be well informed upon, the
subjects you have just named ?*

Certainly ! I deem it to be your imperative duty to
study the subjects well.   The proper management of
children is a vital question,—a mother's question,—and
the most important that can be brought under the consideration of a parent ; and, strange to say, it is one that
has been more neglected than any other.   How many
mothers undertake the responsible management of
children without previous instruction, or without fore-
thought ; they undertake it, as though it may be learned
either by intuition or by instinct, or by affection !   The
consequence is, that frequently they are in a sea of
trouble and uncertainty, tossing about without either rule
or compass : until, too often, their hopes and treasures are
shipwrecked and lost.

The care and management, and consequently the health
and future well-doing of the child, principally devolve
upon the mother ; " for it is the mother after all that has
most to do with the making or marring of the man."*
Dr Guthrie justly remarks that—" Moses might have
never been the man he was unless he had been nursed
by his own mother.   How many celebrated men have
owed their greatness and their goodness to a mother's
training !"   Napoleon owed much to his mother.   " ' The
fate of a child,' said Napoleon, ' is always the work of

---

* *Good Words*, Dr W. Lindsay Alexander, March 1861.

his mother;' and this extraordinary man took pleasure in repeating, that to his mother he owed his elevation. All history confirms this opinion. . . . . The character of the mother influences the children more than that of the father, because it is more exposed to their daily, hourly observation."—*Woman's Mission.*

I am not overstating the importance of the subject in hand ·when I say, that a child is the most valuable treasure in the world, that "he is the precious gift of God," that he is the source of a mother's greatest and purest enjoyment, that he is the strongest bond of affection between her and her husband, and that

> " A babe in a house is a well-spring of pleasure,
> A messenger of peace and love."—*Tupper.*

I have, in the writing of the following pages, had one object constantly in view—namely, health—

> " That salt of life, which does to all a relish give,
> Its standing pleasure, and intrinsic wealth,
> The body's virtue, and the soul's good fortune—health."

If the following pages insist on the importance of one of a mother's duties more than another it is this,—*that the mother herself look well into everything appertaining to the management of her own child.*

Blessed is that mother among mothers of whom it can be said, that "she hath done what she could" for her child—for his welfare, for his happiness, for his health !

For if a mother hath not "done what she could for her child"—mentally, morally, and physically—woe betide the unfortunate little creature;—better had it been for him had he never been born !

ABLUTION.

3. *Is a new-born infant, for the first time, to be washed in warm or in cold water ?*

It is not an uncommon plan to use *cold* water from the first, under the impression of its strengthening the child. This appears to be a cruel and barbarous practice, and is likely to have a contrary tendency. Moreover, it frequently produces either inflammation of the eyes, or stuffing of the nose, or inflammation of the lungs, or looseness of the bowels. Although I do not approve of *cold* water, we ought not to run into an opposite extreme, as *hot* water would weaken and enervate the babe, and thus would predispose him to disease. Luke-warm *rain* water will be the best to wash him with. This, if it be summer, should have its temperature gradually lowered, until it be quite cold ; if it be winter, a *dash* of warm water ought still to be added, to take off the chill.* (By thermometer = 90 to 92 degrees.)

It will be necessary to use soap—Castile soap being the best for the purpose—it being less irritating to the skin than the ordinary soap. Care should be taken that it does not get into the eyes, as it may produce either inflammation or smarting of those organs.

If the skin be delicate, or if there be any excoriation or "breaking-out" on the skin, then glycerine soap, instead of the Castile soap, ought to be used.

4. *At what age do you recommend a mother to commence washing her infant either in the tub, or in the nursery basin ?*

As soon as the navel-string comes away.† Do not be afraid of water,—and that in plenty,—as it is one of the best strengtheners to a child's constitution. How many infants suffer, for the want of water, from excoriation !

5. *Which do you prefer—flannel or sponge—to wash a child with ?*

A piece of flannel is, for the first part of the washing

---

* A nursery-basin (Wedgwood's make, is considered the best), holding either six or eight quarts of water, and which will be sufficiently large to hold the whole body of the child. The basin is generally fitted into a wooden frame which will raise it to a convenient height for the washing of the baby.

† Sir Charles Locock strongly recommends that an infant should be washed *in a tub* from the *very* commencement. He says,—" All those that I superintend *begin* with a tub."—*Letter to the Author.*

very useful—that is to say, to use with the soap, and to loosen the dirt and the perspiration ; but for the finishing-up process, a sponge—a large sponge—is superior to flannel, to wash all away, and to complete the bathing. A sponge cleanses and gets into all the nooks, corners, and crevices of the skin. Besides, sponge, to finish up with, is softer and more agreeable to the tender skin of a babe than flannel. Moreover, a sponge holds more water than flannel, and thus enables you to stream the water more effectually over him. A large sponge will act like a miniature shower bath, and will thus brace and strengthen him.

6. *To prevent a new-born babe from catching cold, is it necessary to wash his head with brandy ?*

It is *not* necessary. The idea that it will prevent cold is erroneous, as the rapid evaporation of heat which the brandy causes is more likely to give than to prevent cold.

7. *Ought that tenacious, paste like substance, adhering to the skin of a new-born babe, to be washed off at the first dressing ?*

It should, provided it be done with a soft sponge and with care. If there be any difficulty in removing the substance, gently rub it, by means of a flannel,* either with a little lard, or fresh butter, or sweet-oil. After the parts have been well smeared and gently rubbed with the lard, or oil, or butter, let all be washed off together, and be thoroughly cleansed away, by means of a sponge and soap and warm water, and then, to com-

---

* Mrs Baines (who has written so much and so well on the Management of Children), in a *Letter* to the Author, recommends flannel to be used in the *first* washing of an infant, which flannel ought afterwards to be burned ; and that the sponge should be only used to complete the process, to clear off what the flannel had already loosened. She also recommends that every child should have his own sponge, each of which should have a particular distinguishing mark upon it, as she considers the promiscuous use of the same sponge to be a frequent cause of *ophthalmia* (inflammation of the eyes). The sponges cannot be kept too clean.

plete the process, gently put him for a minute or two in
his tub. If this paste-like substance be allowed to
remain on the skin, it might produce either an excoriation, or a " breaking-out." Besides, it is impossible, if
that tenacious substance be allowed to remain on it, for
the skin to perform its proper functions.

8. *Have you any general observations to make on the
washing of a new-born infant?*

A babe ought, every morning of his life, to be
thoroughly washed from head to foot ; and this can only
be properly done by putting him bodily either into a tub
or into a bath, or into a large nursery-basin, half filled
with water. The head, before placing him in the bath,
should be first wetted (but not dried) ; then immediately
put him into the water, and, with a piece of flannel well
soaked, cleanse his whole body, particularly his arm-pits,
between his thighs, his groins, and his hams ; then take
a large sponge in hand, and allow the water from it,
well filled, to stream all over the body, particularly over
his back and loins. Let this advice be well observed,
and you will find the plan most strengthening to your
child. The skin must, after every bath, be thoroughly
but quickly dried with warm, dry, soft towels, first
enveloping the child in one, and then gently absorbing
the moisture with the towel, not roughly scrubbing and
rubbing his tender skin as though a horse were being
rubbed down.

The ears must, after each ablution, be carefully and
well dried with a soft dry napkin ; inattention to this
advice has sometimes caused a gathering in the ear—a
painful and distressing complaint ; and at other times it
has produced deafness.

Directly after the infant is dried, all the parts that
are at all likely to be chafed ought to be well powdered.
After he is well dried and powdered, the chest, the back,
the bowels, and the limbs should be gently rubbed,
taking care not to expose him unnecessarily during such
friction.

He ought to be partially washed every evening ; indeed

it may be necessary to use a sponge and a little warm water frequently during the day, namely, each time after the bowels have been relieved. *Cleanliness is one of the grand incentives to health,* and therefore cannot be too strongly insisted upon. If more attention were paid to this subject, children would be more exempt from chafings, "breakings-out," and consequent suffering, than they at present are. After the second month, if the babe be delicate, the addition of two handfuls of table-salt to the water he is washed with in the morning will tend to brace and strengthen him.

· With regard to the best powder to dust an infant with, there is nothing better for general use than starch—the old-fashioned starch *made of wheaten flour*—reduced by means of a pestle and mortar to a fine powder; or Violet Powder, which is nothing more than finely powdered starch scented, and which may be procured of any respectable chemist. Some others are in the habit of using white lead ; but as this is a poison, it ought *on no account* to be resorted to.

9. *If the parts about the groin and fundament be excoriated, what is then the best application ?*

After sponging the parts with tepid *rain* water, holding him over his tub, and allowing the water from a well-filled sponge to stream over the parts, and then drying them with a soft napkin (not rubbing, but gently dabbing with the napkin), there is nothing better than dusting the parts frequently with finely powdered Native Carbonate of Zinc-Calamine Powder. The best way of using this powder is, tying up a little of it in a piece of muslin, and then gently dabbing the parts with it.

Remember excoriations are generally owing to the want of water,—to the want of an abundance of water. An infant who is every morning well soused and well swilled with water seldom suffers either from excoriations, or from any other of the numerous skin diseases. Cleanliness, then, is the grand preventative of, and the best remedy for excoriations. Naaman the Syrian was

ordered " to wash and be clean," and he was healed,
" and his flesh came again like unto the flesh of a little
child and he was clean." This was, of course, a
miracle ; but how often does water, without any special
intervention, act miraculously both in preventing and in
curing skin diseases !

An infant's clothes, napkins especially, ought never to
be washed with soda ; the washing of napkins with
soda is apt to produce excoriations and breakings-out.
" As washerwomen often deny that they use soda, it can
be easily detected by simply soaking a clean white
napkin in fresh water and then tasting the water ; if it
be brackish and salt, soda has been employed."*

10. *Who is the proper person to wash and dress the
babe?*

The monthly nurse, as long as she is in attendance ;
but afterwards the mother, unless she should happen to
have an experienced, sensible, thoughtful nurse, which,
unfortunately, is seldom the case.†

11. *What is the best kind of apron for a mother, or
for a nurse, to wear, while washing the infant?*

Flannel—a good, thick, soft flannel, usually called
bathcoating—apron, made long and full, and which of
course ought to be well dried every time before it is
used.

12. *Perhaps you will kindly recapitulate, and give
me further advice on the subject of the ablution of my
babe.*

Let him by all means, then, as soon as the navel-string

---

* Communicated by Sir Charles Locock to the Author.
† " The Princess of Wales might have been seen on Thursday
taking an airing in a brougham in Hyde Park with her baby—
the future King of England—on her lap, without a nurse, and
accompanied only by Mrs Bruce. The Princess seems a very
pattern of mothers, and it is whispered among the ladies of the
Court that every evening the mother of this young gentleman
may be seen in a flannel dress, in order that she may properly
wash and put on baby's night clothes, and see him safely in bed.
It is a pretty subject for a picture."—*Pall Mall Gazette.*

has separated from the body, be bathed either *in* his tub, or *in* his bath, or *in* his large nursery-basin; for if he is to be strong and hearty, *in* the water every morning he must go. The water ought to be slightly warmer than new milk. It is dangerous for him to remain for a long period in his bath; this, of course, holds good in a tenfold degree if the child have either a cold or pain in his bowels. Take care that, immediately after he comes out of his tub, he is well dried with warm towels. It is well to let him have his bath the first thing in the morning, and before he has been put to the breast; let him be washed before he has his breakfast; it will refresh him and give him an appetite. Besides, he ought to have his morning ablution on an empty stomach, or it may interfere with digestion, and might produce sickness and pain. In putting him in his tub, let his head be the first part washed. We all know, that in bathing in the sea, how much better we can bear the water if we first wet our head; if we do not do so, we feel shivering and starved and miserable. Let there be no dawdling in the washing; let it be quickly over. When he is thoroughly dried with warm *dry* towels, let him be well rubbed with the warm hand of the mother or of the nurse. As I previously recommended, while drying him and while rubbing him, let him repose and kick and stretch either on the warm flannel apron, or else on a small blanket placed on the lap. One bathing in the tub, and that in the morning, is sufficient, and better than night and morning. During the day, as I before observed, he may, after the action either of his bowels or of his bladder, require several spongings of lukewarm water, *for cleanliness is a grand incentive to health and comeliness.*

Remember it is absolutely necessary to every child from his earliest babyhood to have a bath, to be immersed every morning of his life in the water. This advice, unless in cases of severe illness, admits of no exception. Water to the body—to the whole body—is a necessity of life, of health, and of happiness; it wards off disease, it braces the nerves, it hardens the frame, it is the finest

tonic in the world. Oh, if every mother would follow
to the very letter this counsel how much misery, how
much ill-health might then be averted !

MANAGEMENT OF THE NAVEL.

1?. *Should the navel-string be wrapped in* SINGED *rag ?*
There is nothing better than a piece of fine old linen
rag, *unsinged;* when singed, it frequently irritates the
infant's skin.

14. *How ought the navel-string to be wrapped in the
rag ?*
Take a piece of soft linen rag, about three inches wide
and four inches long, and wrap it neatly round the navel
string, in the same manner you would around a cut finger,
and then, to keep on the rag, tie it with a few rounds of
whity-brown thread. The navel-string thus covered
should, pointing upwards, be placed on the belly of the
child, and must be secured in its place by means of a
flannel belly-band.

15. *If after the navel-string has been secured, bleeding
should (in the absence of the medical man) occur, how
must it be restrained ?*
The nurse or the attendant ought immediately to take
off the rag, and tightly, with a ligature composed of four
or five whity-brown threads, retie the navel-string ; and
to make assurance doubly sure, after once tying it, she
should pass the threads a second time around the navel-
string, and tie it again ; and after carefully ascertaining
that it no longer bleeds, fasten it up in the rag as before.
Bleeding of the navel-string rarely occurs, yet, if it should
do so—the medical man not being at hand—the child's
after-health, or even his life, may, if the above directions
be not adopted, be endangered.

16. *When does the navel-string separate from the
child ?*
From five days to a week after birth ; in some cases
not until ten days or a fortnight, or even, in rare cases,
not until three weeks.

17. *If the navel-string does not at the end of a week come away, ought any means to be used to cause the separation?*

Certainly not ; it ought always to be allowed to drop off, which, when in a fit state, it will readily do. Meddling with the navel-string has frequently cost the babe a great deal of suffering, and in some cases even his life.

18. *The navel is sometimes a little sore, after the navel-string comes away; what ought then to be done?*

A little simple cerate should be spread on lint, and be applied every morning to the part affected ; and a white-bread poultice, every night, until it be quite healed.

NAVEL RUPTURE—GROIN RUPTURE.

19. *What are the causes of a rupture of the navel? What ought to be done? Can it be cured?*

(1.) A rupture of the navel is sometimes occasioned by a meddlesome nurse. She is very anxious to cause the navel-string to separate from the infant's body, more especially when it is longer in coming away than usual. She, therefore, before it is in a fit state to drop off, forces it away. (2.) The rupture, at another time, is occasioned by the child incessantly crying. A mother, then, should always bear in mind, that a rupture of the navel is often caused by much crying, and that it occasions much crying ; indeed, it is a frequent cause of incessant crying. A child, therefore, who, without any assignable cause, is constantly crying, should have his navel carefully examined.

A rupture of the navel ought always to be treated early—the earlier the better. Ruptures of the navel can only be *cured* in infancy and in childhood. If it be allowed to run on until adult age, a *cure* is impossible. Palliative means can then only be adopted.

The best treatment is a Burgundy pitch plaster, spread on a soft piece of wash-leather, about the sire of the top of a tumbler, with a properly-adjusted pad (made from the plaster) fastened on the centre of the plaster, which will effectually keep up the rupture, and in a few weeks will cure it. It will be necessary, from time to time, to renew the plaster until the cure be effected. These plasters will be found both more efficacious and pleasant than either truss or bandage ; which latter appliances sometimes gall, and do more harm than they do good.

20. *If an infant have a groin-rupture (an inguinal rupture), can that also be cured ?*

Certainly, if, soon after birth, it be properly attended to. Consult a medical man, and he will supply you with a well-fitting truss, *which will eventually cure him.* If the truss be properly made (under the direction of an experienced surgeon) by a skilful surgical-instrument maker, a beautiful, nicely-fitting truss will be supplied, which will take the proper and exact curve of the lower part of the infant's belly, and will thus keep on without using any under-strap whatever—a great desideratum, as these under-straps are so constantly wetted and soiled as to endanger the patient constantly catching cold. But if. this under-strap is to be superseded, the truss must be made exactly to fit the child—to fit him like a ribbon ; which is a difficult thing to accomplish unless it be fashioned by a skilful workman. It is only lately that these trusses have been made without under-straps. Formerly the under-straps were indispensable necessaries.

These groin-ruptures require great attention and supervision, as the rupture (the bowel) must, before putting on the truss be cautiously and thoroughly returned into the belly ; and much care should be used to prevent the chafing and galling of the tender skin of the babe, which an ill-fitting truss would be sure to occasion. But if care and skill be bestowed on the case, a perfect cure might in due time be ensured. The truss must not be discontinued until a *perfect* cure be effected.

Let me strongly urge you to see that my advice is carried out to the very letter, as a groin-rupture can only be *cured* in infancy and in childhood. If it be allowed to run on, unattended to, until adult age, he will be obliged to wear a truss *all his life*, which would be a great annoyance and a perpetual irritation to him.

CLOTHING.

21. *Is it necessary to have a flannel cap in readiness to put on as soon as the babe is born?*

Sir Charles Locock considers that a flannel cap is *not* necessary, and asserts that all his best nurses have long discarded flannel caps. Sir Charles states that since the discontinuance of flannel caps infants have not been more liable to inflammation of the eyes. Such authority is, in my opinion, conclusive. My advice, therefore, to you is, discontinue by all means the use of flannel caps.

22. *What kind of a belly-band do you recommend—a flannel or a calico one?*

I prefer flannel, for two reasons—first, on account of its keeping the child's bowels comfortably warm; and secondly, because of its not chilling him (and thus endangering cold, &c.) when he wets himself. The belly-band ought to be moderately, but not tightly applied, as, if tightly applied, it would interfere with the necessary movement of the bowels.

23. *When should the belly-band be discontinued?*

When the child is two or three months old. The best way of leaving it off is to tear a strip off daily for a few mornings, and then to leave it off altogether. "Nurses who take charge of an infant when the monthly nurse leaves, are frequently in the habit of at once leaving off the belly-band, which often leads to ruptures, when the child cries or strains. It is far wiser to retain it too long than too short a time; and when a child catches whooping-cough, whilst still very young, it is safer to resume the belly-band.*

24. *Have you any remarks to make on the clothing of an infant.*

A babe's clothing ought to be light, warm, loose, and

---

* Communicated by Sir Charles Locock to the Author.

free from pins. (1.) *It should be light,* without being too airy. Many infant's clothes are both too long and too cumbersome. It is really painful to see how some poor little babies are weighed down with a weight of clothes. They may be said to "bear the burden," and that a heavy one, from the very commencement of their lives! How absurd, too, the practice of making them wear *long* clothes. Clothes to cover a child's feet, and even a little beyond, may be desirable; but for clothes, when the infant is carried about, to reach to the ground, is foolish and cruel in the extreme. I have seen a delicate baby almost ready to faint under the infliction. (2.) *It should be warm,* without being too warm. The parts that ought to be kept warm are the chest, the bowels, and the feet. If the infant be delicate, especially if he be subject to inflammation of the lungs, he ought to wear a fine flannel, instead of his usual shirts, which should be changed as frequently. (3.) *The dress should be loose,* so as to prevent any pressure upon the blood-vessels, which would otherwise impede the circulation, and thus hinder a proper development of the parts. It ought to be loose about the chest and waist, so that the lungs and the heart may have free play. It should be loose about the stomach, so that digestion may not be impeded; it ought to be loose about the bowels, in order that the spiral motion of the intestines may not be interfered with—hence the importance of putting on a belly-band moderately slack; it should be loose about the sleeves, so that the blood may course, without let or hindrance, through the arteries and veins; it ought to be loose, then, everywhere, for nature delights in freedom from restraint, and will resent, sooner or later, any interference. Oh, that a mother would take common sense, and not custom, as her guide! (4.) *As few pins* should be used in the dressing of a baby as possible. Inattention to this advice has caused many a little sufferer to be thrown into convulsions.

The generality of mothers use no pins in the dressing of their children; they tack every part that requires fastening with a needle and thread. They do not even use pins to fasten the baby's·diapers. They make the diapers with loops and tapes, and thus altogether supersede the use of pins in the dressing of an infant. The plan is a good one, takes very little extra time, and deserves to be universally adopted. If pins be used for the diapers, they ought to be the Patent Safety Pins.

25. *Is there any necessity for a nurse being particular in airing an infant's clothes before they are put on ? If she were less particular, would it not make him more hardy ?*

A nurse cannot be too particular on this head. A babe's clothes ought to be well aired the day before they are put on, as they should *not* be put on warm from the fire. It is well, where it can be done, to let him have clean clothes daily. Where this cannot be afforded, the clothes, as soon as they are taken off at night, ought to be well aired, so as to free them from the perspiration, and that they may be ready to put on the following morning. It is truly nonsensical to endeavour to harden a child, or any one else, by putting on damp clothes !

26. *What is your opinion of caps for an infant ?*

The head ought to be kept cool; caps, therefore, are unnecessary. If caps be used at all, they should only be worn for the first month in summer, or for the first two or three months in winter. If a babe take to caps, it requires care in leaving them off, or he will catch cold. When you are about discontinuing them, put a thinner and a thinner one on, every time they are changed, until you leave them off altogether.

But remember, my opinion is, that a child is better *without* caps; they only heat his head, cause undue perspiration, and thus make him more liable to catch cold.

If a babe does not wear a cap in the day, it is not at all necessary that he should wear one at night. He will sleep more comfortably without one, and it will be better for his health. Moreover, night caps injure both the thickness and beauty of the hair

27. *Have you any remarks to make on the clothing of an infant, when, in the winter time, he is sent out for exercise ?*

Be sure that he is well wrapped up. He ought to have under his cloak a knitted worsted spencer, which should button behind ; and if the weather be very cold, a shawl over all; and, provided it be dry above, and the wind be not in the east or in the north-east, he may then brave the weather. He will then come from his walk refreshed and strengthened, for cold air is an invigorating tonic. In a subsequent Conversation, I will indicate the proper age at which a child should be first sent out to take exercise in the open air.

28. *At what age ought an infant " to be shortened ? "*

This, of course, will depend upon the season. In the summer, the right time " for shortening a babe," as it is called, is at the end of two months; in the winter, at the end of three months. But if the right time for "shortening" a child should happen to be in the spring, let it be deferred until the end of May. The English springs are very trying and treacherous; and sometimes, in April, the weather is almost as cold, and the wind as biting as in winter. It is treacherous, for the sun is hot, and the wind, which is at this time of the year frequently easterly, is keen and cutting. I should far prefer "to shorten" a child in the winter than in the early spring.

DIET.

29. *Are you an advocate for putting a baby to the breast soon after birth, or for waiting, as many do, until the third day ?*

The infant ought to be put to the bosom soon after birth : the interest, both of the mother and of the child demands it. It will be advisable to wait three or four hours, that the mother may recover from her fatigue, and, then, the babe must be put to the breast. If this be done, he will generally take the nipple with avidity

It might be said, at so early a period that there is no milk in the bosom; but such is not usually the case. There generally is a *little* from the very beginning, which acts on the baby's bowels like a dose of purgative medicine, and appears to be intended by nature to cleanse the system. But, provided there be no milk at first, the very act of sucking not only gives the child a notion, but, at the same time, causes a draught (as it is usually called) in the breast, and enables the milk to flow easily.

Of course, if there be *no* milk in the bosom—the babe having been applied once or twice to determine the fact—then you must wait for a few hours before applying him again to the nipple, that is to say, until the milk be secreted.

An infant, who, for two or three days, is kept from the breast, and who is fed upon gruel, generally becomes feeble, and frequently, at the end of that time, will not take the nipple at all. Besides, there is a thick cream (similar to the biestings of a cow), which, if not drawn out by the child, may cause inflammation and gathering of the bosom, and, consequently, great suffering to the mother. Moreover, placing him *early* to the breast, moderates the severity of the mother's after pains, and lessens the risk of her flooding. A new-born babe must *not* have gruel given to him, as it disorders the bowels, causes a disinclination to suck, and thus makes him feeble.

30. *If an infant show any disinclination to suck, or if he appear unable to apply his tongue to the nipple, what ought to be done?*

Immediately call the attention of the medical man to the fact, in order that he may ascertain whether he be tongue-tied. If he be, the simple operation of dividing the bridle of the tongue will remedy the defect, and will cause him to take the nipple with ease and comfort.

31. *Provided there be not milk AT FIRST, what ought then to be done?*

Wait with patience; the child (if the mother have no

B

milk) will not, for at least twelve hours, require artificial food. In the generality of instances, then, artificial food is not at all necessary ; but if it should be needed, one-third of new milk and two-thirds of warm water, slightly sweetened with loaf sugar (or with brown sugar, if the babe's bowels have not been opened), should be given, in small quantities at a time, every four hours, until the milk be secreted, and then it must be discontinued.   The infant ought to be put to the nipple every four ·hours, but not oftener, until he be able to find nourishment.

If after the application of the child for a few times, he is unable to find nourishment, then it will be necessary to wait until the milk be secreted.   As soon as it is secreted, he must be applied with great regularity, *alternately* to each breast.

I say *alternately* to each breast. *This is most important advice.* Sometimes a child, for some inexplicable reason, prefers one breast to the other, and the mother, to save a little contention, concedes the point, and allows him to have his own way.   And what is frequently the consequence ?—a gathered breast !

We frequently hear of a babe having no notion of sucking.   This " no notion " may generally be traced to bad management, to stuffing him with food, and thus giving him a disinclination to take the nipple at all.

32. *How often should a mother suckle her infant ?*

A mother generally suckles her baby too often, having him almost constantly at the breast.   This practice is injurious both to parent and to child.   The stomach requires repose as much as any other part of the body ; and how can it have if it be constantly loaded with breast-milk ?   For the first month, he ought to be suckled about every hour and a half ; for the second month, every two hours,—gradually increasing, as he becomes older, the distance of time between, until at length he has it about every four hours.

If a baby were suckled at stated periods, he would only look for the bosom at those times, and be satisfied. A mother is frequently in the habit of giving the child

the breast every time he cries, regardless of the cause.
The cause too frequently is that he has been too often
suckled—his stomach has been overloaded; the little
fellow is consequently in pain, and he gives utterance to
it by cries. How absurd is such a practice ! We may
as well endeavour to put out a fire by feeding it with
fuel. An infant ought to be accustomed to regularity in
everything, in times for sucking, for sleeping, &c. No
children thrive so well as those who are thus early
taught.

33. *Where the mother is* MODERATELY *strong, do you
advise that the infant should have any other food than the
breast ?*

Artificial food must not, for the first five or six
months, be given, if the parent be *moderately* strong ,
of course, if she be feeble, a *little* food will be necessary.
Many delicate women enjoy better health whilst suckling
than at any other period of their lives.

It may be well, where artificial food, in addition to
the mother's own milk, is needed, and before giving any
farinaceous food whatever (for farinaceous food until a
child is six or seven months old is injurious), to give,
through a feeding-bottle, every night and morning, in
addition to the mother's breast of milk, the following
Milk-Water-and-Sugar-of-Milk Food :—

> Fresh milk, from ONE cow ;
> Warm water, of each a quarter of a pint,
> Sugar-of-milk one tea-spoonful.

The sugar-of-milk should first be dissolved in the warm
water, and then the fresh milk *unboiled* should be mixed
with it. The sweetening of the above food with sugar-
of-milk, instead of with lump sugar, makes the food
more to resemble the mother's own milk. The infant
will not, probably, at first take more than half of the
above quantity at a time, even if he does so much as
that ; but still the above are the proper proportions ; and
as he grows older, he will require the whole of it at a
meal.

34. *What food, when a babe is six or seven months old, is the best substitute for a mother's milk?*

The food that suits one infant will not agree with another. (1.) The one that I have found the most generally useful, is made as follows :—Boil the crumb of bread for two hours in water, taking particular care that it does not burn ; then add only a *little* lump-sugar (or *brown* sugar, if the bowels be costive), to make it palatable. When he is six or seven months old, mix a little new milk—the milk of ONE cow—with it gradually as he becomes older, increasing the quantity until it be nearly all milk, there being only enough water to boil the bread ; the milk should be poured boiling hot on the bread. Sometimes the two milks—the mother's and the cow's milk—do not agree ; when such is the case, let the milk be left out, both in this and in the foods following, and let the food be made with water, instead of with milk and water. In other respects, until the child is weaned, let it be made as above directed ; when he is weaned, good fresh cow's milk MUST, as previously recommended, be used. (2.) Or cut thin slices of bread into a basin, cover the bread with *cold* water, place it in an oven for two hours to bake ; take it out, beat the bread up with a fork, and then slightly sweeten it. This is an excellent food. (3.) If the above should not agree with the infant (although, if properly made, they almost invariably do), "tous-les-mois" may be given.* (4.) Or Robb's Biscuits, as it is "among the best bread compounds made out of wheat-flour, and is almost always readily digested."—*Routh.*

(5.) Another good food is the following :—Take about a pound of flour, put it in a cloth, tie it up tightly, place it in a saucepanful of water, and let it boil for four or five hours ; then take it out, peel off the outer rind, and

---

* "Tous-les-mois" is the starch obtained from the tuberous roots of various species of *canna*, and is imported from the West Indies. It is very similar to arrow-root. I suppose it is called "tuos-les-mois," as it is good to be eaten all the year round.

the inside will be found quite dry, which grate. (6.) Another way of preparing an infant's food, is to bake flour—biscuit flour—in a slow oven, until it be of a light fawn colour. Baked flour ought, after it is baked, to be reduced, by means of a rolling-pin, to a fine powder, and should then be kept in a covered tin, ready for use. (7.) An excellent food for a baby is baked crumbs of bread. The manner of preparing it is as follows:— Crumb some bread on a plate; put it a little distance from the fire to dry. When dry, rub the crumbs in a mortar, and reduce them to a fine powder; then pass them through a sieve. Having done which, put the crumbs of bread into a slow oven, and let them bake until they be of a light fawn colour. A small quantity either of the boiled, or of the baked flour, or of the baked crumb of bread, ought to be made into food, in the same way as gruel is made, and should then be slightly sweetened, according to the state of the bowels, either with lump or with brown sugar.

(8.) Baked flour sometimes produces constipation; when such is the case, Mr Appleton, of Budleigh Salterton, Devon, wisely recommends a mixture of baked flour, and prepared oatmeal,* in the proportion of two of the former and one of the latter. He says:—"To avoid the constipating effects, I have always had mixed, before baking, one part of prepared oatmeal with two parts of flour; this compound I have found both nourishing, and regulating to the bowels. One table-spoonful of it, mixed with a quarter of a pint of milk, or milk and water, when well boiled, flavoured and sweetened with white sugar, produces a thick, nourishing, and delicious food for infants or invalids." He goes on to remark:—"I know of no food, after repeated trials, that can be so strongly recommended by the profession to all mothers in the rearing of their infants, without or with the aid of the breasts, at the same time relieving them of much

---

* If there is any difficulty in obtaining *prepared* oatmeal, Robinson's Scotch Oatmeal will answer equally as well.

draining and dragging whilst nursing with an insufficiency of milk, as baked flour and oatmeal.*

(9.) A ninth food may be made with "Farinaceous Food for Infants, prepared by Hards of Dartford." If Hard's Farinaceous food produces costiveness—as it sometimes does—let it be mixed either with equal parts or with one-third of Robinson's Scotch Oatmeal. The mixture of the two together makes a splendid food for a baby. (10.) A tenth, and an excellent one, may be made with rusks, boiled for an hour in water, which ought then to be well beaten up, by means of a fork, and slightly sweetened with lump sugar. Great care should be taken to select good rusks, as few articles vary so much in quality. (11.) An eleventh is—the top crust of a baker's loaf, boiled for an hour in water, and then moderately sweetened with lump sugar. If, at any time, the child's bowels should be costive, *raw* must be substituted for *lump* sugar. (12.) Another capital food for an infant is that made by Lemann's Biscuit Powder.† (13.) Or, Brown and Polson's Patent Corn Flour will be found suitable. Francatelli, the Queen's cook, in his recent valuable work, gives the following formula for making it:—"To one dessert-spoonful of Brown and Polson, mixed with a wineglassful of cold water, add half a pint of boiling water; stir over the fire for five minutes; sweeten lightly, and feed the baby; but if the infant is being brought up by the hand, this food should then be mixed with milk—not otherwise." (14.) A fourteenth is Neaves' Farinaceous Food for Infants, which is a really good article of diet for a babe; it is not so binding to the bowels as many of the farinaceous foods are, which is a great recommendation.

(15.) The following is a good and nourishing food for

---

* *British Medical Journal,* Dec. 18, 1858.

† Lemann's Biscuit Powder cannot be too strongly recommended :—It is of the finest quality, and may be obtained of Lemann, Threadneedle Street, London. An extended and an extensive experience confirms me still more in the good opinion I have of this food.

a baby :—Soak for an hour, some *best* rice in cold water; strain, and add fresh water to the rice; then let it simmer till it will pulp through a sieve ; put the pulp and the water in a saucepan, with a lump or two of sugar, and again let it simmer for a quarter of an hour ; a portion of this should be mixed with one-third of fresh milk, so as to make it of the consistence of good cream. This is an excellent food for weak bowels.

When the baby is six or seven months old, new milk should be added to any of the above articles of food, in a similar way to that recommended for boiled bread.

(16.) For a delicate infant, lentil powder, better known as Du Barry's " Ravalenta Arabica," is invaluable. It ought to be made into food, with new milk, in the same way that arrow-root is made, and should be moderately sweetened with loaf-sugar. Whatever food is selected ought to be given by means of a nursing bottle.

If a child's bowels be relaxed and weak, or if the motions be offensive, the milk *must* be boiled, but not otherwise. The following (17) is a good food when an infant's bowels are weak and relaxed :—" Into five large spoonfuls of the purest water, rub smooth one dessert-spoonful of fine flour. Set over the fire five spoonfuls of new milk, and put two bits of sugar into it; the moment it boils, pour it into the flour and water, and stir it over a slow fire twenty minutes."

Where there is much emaciation, I have found (18) genuine arrow-root* a very valuable article of food for an infant, as it contains a great deal of starch, which starch helps to form fat and to evolve caloric (heat)—both of which a poor emaciated chilly child stands so much in need of. It must be made with equal parts of water and of good fresh milk, and ought to be slightly sweetened with loaf sugar; a small pinch of table salt should be added to it.

---

* Genuine arrow-root, of first-rate quality, and at a reasonable price, may be obtained of H. M. Plumbe, arrow-root merchant, 8 Alie Place, Great Alie Street, Aldgate, London, E.

Arrow-root will not, as milk will, give bone and muscle; but it will give—what is very needful to a delicate child—fat and warmth. Arrow-root, as it is principally composed of starch, comes under the same category as cream, butter, sugar, oil, and fat. Arrowroot, then, should always be given with new milk (mixed with one-half of water); it will then fulfil, to perfection, the exigencies of nourishing, of warming, and fattening the child's body.

New milk, composed in due proportions as it is, of cream and of skim milk—the very acme of perfection—is the only food, *which of itself alone*, will nourish and warm and fatten. It is, for a child, *par excellence*, the food of foods!

Arrow-root, and all other farinaceous foods are, for a child, only supplemental to milk—new milk being, for the young, the staple food of all other kinds of foods whatever.

But bear in mind, *and let there be no mistake about it*, that farinaceous food, be it what it may, until the child be six or seven months old, until, indeed, he *begin* to cut his teeth, is *not* suitable for a child; until then, *The Milk-water-salt-and-sugar Food* (see page 29) is usually, if he be a dry-nursed child, the best artificial food for him.

I have given you a large and well-tried infant's dietary to chose from, as it is sometimes difficult to fix on one that will suit; but, remember, if you find one of the above to agree, keep to it, as a babe requires a simplicity in food—a child a greater variety.

Let me, in this place, insist upon the necessity of great care and attention being observed in the preparation of any of the above articles of diet. A babe's stomach is very delicate, and will revolt at either ill-made, or lumpy, or burnt food. Great care ought to be observed as to the cleanliness of the cooking utensils. The above directions require the strict supervision of the mother.

Broths have been recommended, but, for my own part, I think that, for a *young* infant, they are objectionable;

they are apt to turn acid on the stomach, and to cause flatulence and sickness; they, sometimes, disorder the bowels and induce griping and purging.

Whatever artificial food is used ought to be given by means of a bottle, not only as it is a more natural way than any other of feeding a baby, as it causes him to suck as though he were drawing it from the mother's breasts, but as the act of sucking causes the salivary glands to press out their contents, which materially assist' digestion. Moreover, it seems to satisfy and comfort him more than it otherwise would do.

One of the best, if not *the best* feeding bottle I have yet seen, is that made by Morgan Brothers, 21 Bow Lane, London. It is called " The Anglo-French Feeding Bottle." S. Maw, of 11 Aldersgate Street, London, has also brought out an excellent one—" The Fountain Infant's Feeding Bottle." Another good one is " Mather's Infant's Feeding Bottle." Either of these three will answer the purpose admirably. I cannot speak in terms too highly of these valuable inventions.

The food ought to be of the consistence of good cream, and should be made fresh and fresh. It ought to be given milk-warm. Attention must be paid to the cleanliness of the vessel, and care should be taken that the milk be that of ONE cow,* and that it be new and of good quality; for if not it will turn acid and sour,

---

* I consider it to be of immense importance to the infant, that the milk be had from ONE cow. A writer in the *Medical Times and Gazette*, speaking on this subject, makes the following sensible remarks :—" I do not know if a practice common among French ladies when they do not nurse, has obtained the attention among ourselves which it seems to me to deserve. When the infant is to be fed with cow milk that from various cows is submitted to examination by the medical man, and if possible, tried on some child, and when the milk of any cow has been chosen, no other milk is ever suffered to enter the child's lips, for a French lady would as soon offer to her infant's mouth the breasts of half a dozen wet-nurses in the day, as mix together the milk of various cows, which must differ, even as the animals themselves, in its constituent qualities. Great attention is also paid to the pasture, or other food of the cow thus appropriated."

and disorder the stomach, and will thus cause either flatulence or looseness of the bowels, or perhaps convulsions. The only way to be sure of having it from *one* cow, is (if you have not a cow of your own), to have the milk from a *respectable* cow-keeper, and to have it brought to your house in a can of your own (the London milk-cans being the best for the purpose). The better plan is to have two cans, and to have the milk fresh and fresh every night and morning. The cans, after each time of using, ought to be scalded out; and, once a week the can should be filled with *cold* water, and the water should be allowed to remain in it until the can be again required.

Very little sugar should be used in the food, as much sugar weakens the digestion. A small pinch of table-salt ought to be added to whatever food is given, as "the best savour is salt." Salt is most wholesome—it strengthens and assists digestion, prevents the formation of worms, and, in small quantities, may with advantage be given (if artificial food be used) to the youngest baby.

35. *Where it is found to be absolutely necessary to give an infant artificial food* WHILST SUCKLING, *how often ought he to be fed?*

Not oftener than twice during the twenty-four hours, and then only in *small* quantities at a time, as the stomach requires rest, and at the same time, can manage to digest a little food better than it can a great deal. Let me again urge upon you the importance, if it be at all practicable, of keeping the child *entirely* to the breast for the first five or six months of his existence. Remember there is no *real* substitute for a mother's milk; there is no food so well adapted to his stomach; there is no diet equal to it in developing muscle, in making bone, or in producing that beautiful plump rounded contour of the limbs; there is nothing like a mother's milk *alone* in making a child contented and happy, in laying the foundation of a healthy constitution, in preparing the body for a long life, in giving him tone to resist disease, or in causing him to cut his teeth easily and well; in short, *the mother's milk is the greatest temporal blessing an infant can possess.*

As a general rule, therefore, when the child and the mother are tolerably strong, he is better *without artificial* food until he have attained the age of three or four months; then, it will usually be necessary to feed him with *The Milk-water-and-sugar-of-milk Food* (see p. 19) twice a day, so as gradually to prepare him to be weaned (if possible) at the end of nine months. The food mentioned in the foregoing Conversation will, when he is six or seven months old, be the best for him.

36. *When the mother is not able to suckle her infant herself, what ought to be done?*

It must first be ascertained, *beyond all doubt,* that a mother is not able to suckle her own child. Many delicate ladies do suckle their infants with advantage, not only to their offspring, but to themselves. "I will maintain," says Steele, "that the mother grows stronger by it, and will have her health better than she would have otherwise. She will find it the greatest cure, and preservative for the vapours [nervousness] and future miscarriages, much beyond any other remedy whatsoever. Her children will be like giants, whereas otherwise they are but living shadows, and like unripe fruit; and certainly if a woman is strong enough to bring forth a child, she is beyond all doubt strong enough to nurse it afterwards."

Many mothers are never so well as when they are nursing; besides, suckling prevents a lady from becoming pregnant so frequently as she otherwise would. This, if she be delicate, is an important consideration, and more especially if she be subject to miscarry. The effects of miscarriage are far more weakening than those of suckling.

A hireling, let her be ever so well inclined, can never have the affection and unceasing assiduity of a mother, and, therefore, cannot perform the duties of suckling with equal advantage to the baby.

The number of children who die under five years of age is enormous—many of them from the want of the mother's milk. There is a regular "parental baby-slaughter"—"a massacre of the innocents"—constantly going on in England, in consequence of infants being thus deprived of their proper nutriment and just dues! The mortality from this cause is frightful, chiefly occurring among rich people who are either too grand, or, from luxury, too delicate to perform such duties: poor married women, as a rule, nurse their own children, and, in consequence, reap their reward.

If it be ascertained, *past all doubt*, that a mother cannot suckle her child, then, if the circumstances of the parents will allow—and they ought to strain a point to accomplish it—a healthy wet-nurse should be procured, as, of course, the food which nature has supplied is far, very far superior to any invented by art. Never bring up a baby, then, if you can possibly avoid it, on *artificial* food. Remember, as I proved in a former Conversation, there is in early infancy no *real* substitute for either a mother's or a wet-nurse's milk. It is impossible to imitate the admirable and subtle chemistry of nature. The law of nature is, that a baby, for the first few months of his existence, shall be brought up by the breast; and nature's law cannot be broken with impunity.* It will be imperatively necessary then—

"To give to nature what is nature's due."

Again, in case of a severe illness occurring during the first nine months of a child's life, what a comfort either the mother's or the wet-nurse's milk is to him! it often determines whether he shall live or die. But if a wet-nurse cannot fill the place of a mother, then asses' milk will be found the best substitute, as it approaches nearer, in composition, than any other animal's, to human milk; but it is both difficult and expensive to obtain. The next best substitute is goats' milk. Either the one or

---

* For further reasons why artificial food is not desirable, at an early period of infancy, see answer to 35th question, page 26.

the other ought to be milked fresh and fresh, when wanted, and should be given by means of a feeding-bottle. Asses' milk is more suitable for a *delicate* infant, and goats' milk for a *strong* one.

If neither asses' milk nor goats' milk can be procured, then the following *Milk-water-salt-and-sugar Food*, from the very commencement, should be given; and as I was the author of the formula,* I beg to designate it as—*Pye Chavasse's Milk Food :*—

> New milk, the produce of ONE *healthy* cow;
> Warm water, of each, equal parts;
> Table salt, a few grains—a small pinch;
> Lump sugar, a sufficient quantity, to slightly sweeten it.

The milk itself ought not to be heated over the fire,† but should, as above directed, be warmed by the water; it must, morning and evening, be had fresh and fresh. The milk and water should be of the same temperature as the mother's milk, that is to say, at about ninety degrees Fahrenheit. It ought to be given by means of either Morgan's, or Maw's, or Mather's feeding-bottle,‡ and care must be taken to *scald* the bottle out twice a day, for if attention be not paid to this point, the delicate stomach of an infant is soon disordered. The milk should, as he grows older, be gradually increased and the water decreased, until two-thirds of milk and one-third of water be used; but remember, that either *much* or *little* water must *always* be given with the milk.

The above is my old form, and which I have for many years used with great success. Where the above food does not agree (and no food except a healthy mother's own milk does *invariably* agree) I occasionally substitute

---

* It first appeared in print in the 4th edition of *Advice to a Mother*, 1852.

† It now and then happens, that if the milk be not boiled, the motions of an infant are offensive; *when such is the case*, let the milk be boiled, but not otherwise.

‡ See answer to Question 24, page 24.

sugar-of-milk for the lump sugar, in the proportion of a tea-spoonful of sugar-of-milk to every half-pint of food.

If your child bring up his food, and if the ejected matter be sour-smelling, I should advise you to leave out the sugar-of-milk altogether, and simply to let the child live, for a few days, on milk and water alone, the milk being of *one* cow, and in the proportion of two-thirds to one-third of *warm* water—not *hot* water; the milk should not be scalded with *hot* water, as it injures its properties; besides, it is only necessary to give the child his food with the chill just off. The above food, where the stomach is disordered, is an admirable one, and will often set the child to rights without giving him any medicine whatever. Moreover, there is plenty of nourishment in it to make the babe thrive; for after all it is the milk that is the important ingredient in all the foods of infants; they can live on it, and on it alone, and thrive amazingly.

Mothers sometimes say to me, that farinaceous food makes their babes flatulent, and that my food (*Pye Chavasse's Milk Food*) has not that effect.

The reason of farinaceous food making babes, until they have *commenced* cutting their teeth, " windy " is, that the starch of the farinaceous food (and all farinaceous foods contain more or less of starch) is not digested, and is not, as it ought to be, converted by the saliva into sugar :* hence " wind " is generated, and pain and convulsions often follow in the train.

The great desideratum, in devising an infant's formula for food, is to make it, until he be nine months old, to resemble as much as possible, a mother's own milk ; and which my formula, as nearly as is practicable, does resemble : hence its success and popularity.

As soon as a child begins to cut his teeth the case is altered, and *farinaceous food, with milk and with water*, becomes an absolute necessity

I wish, then, to call your especial attention to the

---

* See Pye Chavasse's *Counsel to a Mother*, 3d edition.

following facts, for they are facts :—Farinaceous foods, *of all kinds,* before a child *commences* cutting his teeth (which is when he is about six or seven months old) are worse than useless—they are, positively, injurious ; they are, during the early period of infant life, perfectly indigestible, and may bring on—which they frequently do —convulsions. A babe fed on farinaceous food alone would certainly die of starvation ; for, " up to six or seven months of age, infants have not the power of digesting farinaceous or fibrinous substances."—Dr Letheby on *Food.*

A babe salivary glands, until he be six or seven months old, does not secrete its proper fluid—namely, ptyalin, and consequently the starch of the farinaceous food—and all farinaceous food contains starch—is not converted into dextrine and grape-sugar, and is, therefore, perfectly indigestible and useless—nay, injurious to an infant, and may bring on pain and convulsions, and even death ; hence, the giving of farinaceous food, until a child be six or seven months old, is one and the principal cause of the frightful infant mortality at the present time existing in England, and which is a disgrace to any civilized land !

In passing, allow me to urge you never to stuff a babe —never to overload his little stomach with food ; it is far more desirable to give him a little not enough, than to give him a little too much. Many a poor child has been, like a young bird, killed with stuffing. If a child be at the breast, and at the breast alone, there is no fear of his taking too much ; but if he be brought up on artificial food, there is great fear of his over-loading his stomach. Stuffing a child brings on vomiting and bowel-complaints, and a host of other diseases which now it would be tedious to enumerate. Let me, then, urge you on no account, to over-load the stomach of a little child.

There will, then, in many cases, be quite sufficient nourishment in the above ; I have known some robust infants brought up on it, and on it alone, without a particle of farinaceous food, or of any other food, in any shape or form whatever. But if it should not agree with the child, or if there should not be sufficient nourishment in it, then the food recommended in answer to No. 34 question ought to be given, with this only difference—a little new milk *must* from the beginning be added, and should be gradually increased, until nearly all milk be used.

The milk, as a general rule, ought to be *unboiled ;* but if it purge violently, or if it cause offensive motions—which it sometimes does—then it must be boiled. The moment the milk boils up, it should be taken off the fire.

Food ought for the first month to be given about every two hours ; for the second month, about every three hours ; lengthening the space of time as the baby advances in age. A mother must be careful not to over-feed a child, as over-feeding is a prolific source of disease.

Let it be thoroughly understood, and let there be no mistake about it, that a babe during the first nine months of his life, MUST have—it is absolutely necessary for his very existence—milk of some kind, as the staple and principal article of his diet, either mother's `t wet-nurse's, or asses', or goats', or cow's milk.

37. *How would you choose a wet-nurse ?*

I would inquire particularly into the state of her health ; whether she be of a healthy family, of a consumptive habit, or if she or any of her family have laboured under " king's evil ; " ascertaining if there be any seams or swellings about her neck ; any eruptions or blotches upon her skin ; if she has a plentiful breast of milk, and if it be of good quality* (which may readily be ascertained by milking a little into a glass) ; if she has good nipples, sufficiently long for the baby to hold ; that they be not sore ; and if her own child be of the same, or nearly of the same age, as the one you wish her

---

* "It should be thin, and of a bluish-white colour, sweet to the taste, and when allowed to stand, should throw up a considerable quantity of cream,"—*Maunsell and Evenson on the Diseases of Children.*

to nurse. Ascertain, whether she menstruate during suckling ; if she does, the milk is not so good and nourishing, and you had better decline taking her.* Assure yourself that her own babe is strong and healthy and that he is free from a sore mouth, and from a " breaking-out " of the skin. Indeed, if it be possible to procure such a wet-nurse, she ought to be from the country, of ruddy complexion, of clear skin, and of between twenty and five-and-twenty years of age, as the milk will then be fresh, pure, and nourishing.

I consider it to be of great importance that the infant of the wet-nurse should be, as nearly as possible, of the same age as your own, as the milk varies in quality according to the age of the child. For instance, during the commencement of suckling, the milk is thick and creamy, similar to the biestings of a cow, which, if given to a babe of a few months old, would cause derangement of the stomach and bowels. After the first few days, the appearance of the milk changes ; it becomes of a bluish-white colour, and contains less nourishment. The milk gradually becomes more and more nourishing as the infant becomes older and requires more support.

In selecting a wet-nurse for a very small and feeble babe, you must carefully ascertain that the nipples of the wet-nurse are good and soft, and yet not very large. If they be very large, the child's mouth being very small, he may not be able to hold them. You must note, too, whether the milk flows readily from the nipple into the child's mouth ; if it does not, he may not have strength to draw it, and he would soon die of starvation. The only way of ascertaining whether the infant really draws the milk from the nipple, can be done by examining the mouth of the child *immediately* after his taking the

---

* Sir Charles Locock considers that a woman who menstruates during lactation is objectionable as a wet-nurse, and "that as a mother with her first child is more liable to that objection, that a second or third child's mother is more eligible than a first."—*Letter to the Author.*

C

breast, and seeing for yourself whether there be actually milk, or not, in his mouth.

Very feeble new-born babes sometimes cannot take the bosom, be the nipples and the breasts ever so good, and although Maw's nipple-shield and glass tube had been tried. In such a case, cow's milk-water-sugar-and-salt, as recommended at page 29, must be given in small quantities at a time—from two to four tea-spoonfuls—but frequently; if the child be awake, every hour, or every half hour, both night and day, until he be able to take the breast. If, then, a puny, feeble babe is only able to take but little at a time, and that little by tea-spoonfuls, he must have little and often, in order that " many a little might make a mickle."

I have known many puny, delicate children who had not strength to hold the nipple in their mouths, but who could take milk and water (as above recommended) by tea-spoonfuls only at a time, with steady perseverance, and giving it every half hour or hour (according to the quantity swallowed), at length be able to take the breast, and eventually become strong and hearty children; but such cases require unwearied watching, perseverance, and care. Bear in mind, then, that the smaller the quantity of the milk and water given at a time, the oftener must it be administered, as, of course, the babe must have a certain quantity of food to sustain life.

38. *What ought to be the diet either of a wet-nurse, or of a mother, who is suckling?*

It is a common practice to cram a wet-nurse with food, and to give her strong ale to drink, to make good nourishment and plentiful milk! This practice is absurd; for it either, by making the nurse feverish, makes the milk more sparing than usual, or it causes the milk to be gross and unwholesome. On the other hand, we must not run into an opposite extreme. The mother, or the wet-nurse, by using those means most conducive to her own health, will best advance the interest of her little charge.

A wet-nurse, ought to live somewnat in the following way :—Let her for breakfast have black tea, with one or two slices of cold meat, if her appetite demand it, but not otherwise. It is customary for a wet-nurse to make a hearty luncheon ; of this I do not approve. If she feel either faint or low at eleven o'clock, let her have either a tumbler of porter, or of mild fresh ale, with a piece of dry toast soaked in it. She ought not to dine later than half-past one or two o'clock ; she should eat, for dinner, either mutton or beef, with either mealy potatoes, or asparagus, or French beans, or secale, or turnips, or broccoli, or cauliflower, and stale bread. Rich pastry, soups, gravies, high-seasoned dishes, salted meats, greens, and cabbage, must one and all be carefully avoided ; as they only tend to disorder the stomach, and thus to deteriorate the milk.

It is a common remark, that "a mother who is suckling may eat anything." I do not agree with this opinion. Can impure or improper food make pure and proper milk, or can impure and improper milk make good blood for an infant, and thus good health ?

The wet-nurse ought to take with her dinner a moderate quantity of either sound porter, or of mild (but not old or strong) ale. Tea should be taken at half past five or six o'clock ; supper at nine, which should consist either of a slice or two of cold meat, or of cheese if she prefer it, with half a pint of porter or of mild ale ; occasionally a basin of gruel may with advantage be substituted. Hot and late suppers are prejudicial to the mother, or to the wet-nurse, and, consequently, to the child. The wet-nurse ought to be in bed every night by ten o'clock.

It might be said, that I have been too minute and particular in my rules for a wet-nurse ; but when it is considered of what importance good milk is to the welldoing of an infant, in making him strong and robust, not only now, but as he grows up to manhood, I shall, I trust, be excused for my prolixity.

39. *Have you any more hints to offer with regard to the management of a wet-nurse ?*

A wet-nurse is frequently allowed to remain in bed

until a late hour in the morning, and during the day to continue in the house, as if she were a fixture! How is it possible that any one, under such treatment, can continue healthy? A wet nurse ought to rise early, and, if the weather and season will permit, take a walk, which will give her an appetite for breakfast, and will make a good meal for her little charge. This, of course, cannot, during the winter months, be done ; but even then, she ought, some part of the day, to take every opportunity of walking out ; indeed, in the summer time she should live half the day in the open air.

She ought strictly to avoid crowded rooms ; her mind should be kept calm and unruffled, as nothing disorders the milk so much as passion, and other violent emotions of the mind ; a fretful temper is very injurious, on which account you should, in choosing your wet-nurse, endeavour to procure one of a mild, calm, and placid disposition.*

A wet-nurse ought never to be allowed to dose her little charge either with Godfrey's Cordial, or with Dalby's Carminative, or with Syrup of White Poppies, or with medicine of any kind whatever. Let her thoroughly understand this, and let there be no mistake in the matter. Do not for one moment allow your children's health to be tampered and trifled with. A baby's health is too precious to be doctored, to be experimented upon, and to be ruined by an ignorant person.

40. *Have the goodness to state at what age a child ought to be weaned.*

This, of course, must depend both upon the strength of the child, and upon the health of the parent ; on an

---

* " 'The child is poisoned.'
'Poisoned ! by whom ?'
'By you. You have been fretting.'
'Nay, indeed, mother. How can I help fretting ?
'Don't tell me, Margaret. A nursing mother has no business to fret. She must turn her mind away from her grief to the comfort that lies in her lap. Know you not that the child pines if the mother vexes herself?' "—*The Cloister and the Hearth.* By Charles Reade.

average, nine months is the proper time. If the mother
be delicate, it may be found necessary to wean the infant
at six months; or if he be weak, or labouring under any
disease, it may be well to continue suckling him for
twelve months; but after that time, the breast will do
him more harm than good, and will, moreover, injure the
mother's health, and may, if she be so predisposed, excite
consumption.

41. *How would you recommend a mother to act when
she weans her child?*

She ought, as the word signifies, do it gradually—that
is to say, she should, by degrees, give him less and less
of the breast, and more and more of artificial food; at
length, she must only suckle him at night; and lastly,
it would be well for the mother either to send him away,
or to leave him at home, and, for a few days, to go away
herself.

A good plan is, for the nurse-maid to have a half-pint
bottle of new milk—which has been previously boiled*
—in the bed, so as to give a little to him in lieu of the
breast. The warmth of the body will keep the milk of
a proper temperature, and will supersede the use of
lamps, of candle-frames, and of other troublesome
contrivances.

42. *While a mother is weaning her infant, and after
she have weaned him, what ought to be his diet?*

Any one of the foods recommended in answer to question 34, page 20.

43. *If a child be suffering severely from "wind," is
there any objection to the addition of a small quantity
either of gin or of peppermint to his food to disperse it?*

It is a murderous practice to add either gin or peppermint of the shops (which is oil of peppermint dissolved
in spirits) to his food. Many children have, by such a
practice, been made puny and delicate, and have gradually
dropped into an untimely grave. An infant who is kept,

---

* The previous boiling of the milk will prevent the warmth of
the bed turning the milk sour, which it otherwise would do.

for the first five or six months, *entirely* to the breast
—more especially if the mother be careful in her own
diet—seldom suffers from " wind ; " those, on the contrary, who have much or improper food,* suffer severely.

Care in feeding, then, is the grand preventative of
" wind ;" but if, notwithstanding all your precautions,
the child be troubled with flatulence, the remedies recommended under the head of Flatulence will generally
answer the purpose.

44. *Have you any remarks to make on sugar for sweetening a baby's food ?*

A *small* quantity of sugar in an infant's food is
requisite, sugar being nourishing and fattening, and making cow's milk to resemble somewhat in its properties
human milk ; but, bear in mind, *it must be used sparingly.*
Much sugar cloys the stomach, weakens the digestion,
produces acidity, sour belchings, and wind :—

> "Things sweet to taste, prove in digestion sour.'
> *Shakspeare.*

If a babe's bowels be either regular or relaxed, *lump*
sugar is the best for the purpose of sweetening his food ;
if his bowels are inclined to be costive, *raw* sugar ought
to be substituted for lump sugar, as *raw* sugar acts on a
young babe as an aperient, and, in the generality of
cases, is far preferable to physicking him with opening
medicine. An infant's bowels, whenever it be practicable (and it generally is), ought to be regulated by a
judicious dietary rather than by physic.

VACCINATION AND RE-VACCINATION.

45. *Are you an advocate for vaccination ?*

Certainly. I consider it to be one of the greatest
blessings ever conferred upon mankind. Small-pox,

---

* For the first five or six months never, if you can possibly
avoid it, give artificial food to an infant who is sucking. There
is nothing, in the generality of cases, that agrees, for the first
few months, like the mother's milk *alone*

before vaccination was adopted, ravaged the country like a plague, and carried off thousands annually ; and those who did escape with their lives were frequently made loathsome and disgusting objects by it. Even inoculation (which is cutting for the small-pox) was attended with danger, more especially to the unprotected—as it caused the disease to spread like wildfire, and thus it carried off immense numbers.

Vaccination is one, and an important cause of our increasing population ; small-pox, in olden times, decimated the country.

46. *But vaccination does not always protect a child from small-pox ?*

I grant you that it does not *always* protect him, *neither does inoculation ;* but when he is vaccinated, if he take the infection, he is seldom pitted, and very rarely dies, and the disease assumes a comparatively mild form. There are a few, very few fatal cases recorded after vaccination, and these may be considered as only exceptions to the general rule ; and, possibly, some of these may be traced to the arm, when the child was vaccinated, not having taken proper effect.

If children, and adults were *re-vaccinated,*—say every seven years after the first vaccination,—depend upon it, even these rare cases would not occur, and in a short time small-pox would be known only by name.

47. *Do you consider it, then, the imperative duty of a mother, in every case, to have, after the lapse of every seven years, her children re-vaccinated ?*

I decidedly do : it would be an excellent plan for *every* person, once every seven years to be re-vaccinated, and even oftener, if small-pox be rife in the neighbourhood. Vaccination, however frequently performed, can never do the slightest harm, and might do inestimable good. Small-pox is both a pest and a disgrace, and ought to be constantly fought and battled with, until it be banished (which it may readily be) the kingdom.

I say that small-pox is a pest ; it is worse than the plague, for if not kept in subjection, it is more general—

sparing neither young nor old, rich nor poor, and commits greater ravages than the plague ever did. Smallpox is a disgrace : it is a disgrace to any civilised land, as there is no necessity for its presence : if cow-pox were properly and frequently performed, small-pox would be unknown. Cow-pox is a weapon to conquer small-pox and to drive it ignominiously from the field.

My firm belief, then, is, that if *every* person were, *every seven years*, duly and properly vaccinated, small-pox might be utterly exterminated ; but as long as there are such lax notions on the subject, and such gross negligence, the disease will always be rampant, for the poison of small-pox never slumbers nor sleeps, but requires the utmost diligence to eradicate it. The great Dr Jenner, the discoverer of cow-pox as a preventative of small-pox, strongly advocated the absolute necessity of *every* person being re-vaccinated once every seven years, or even oftener, if there was an epidemic of small-pox in the neighbourhood.

48. *Are you not likely to catch not only the cow-pox, but any other disease that the child has from whom the matter is taken ?*

The same objection holds good in cutting for small pox (inoculation)—only in a ten-fold degree—small-pox being such a disgusting complaint. Inoculated smallpox frequently produced and· left behind inveterate " breakings-out," scars, cicatrices, and indentations of the skin, sore eyes, blindness, loss of eyelashes, scrofula, deafness—indeed, a long catalogue of loathsome diseases. A medical man, of course, will be careful to take the cow-pox matter from a healthy child.

49. *Would it not be well to take the matter direct from the cow ?*

If a doctor be careful—which, of course, he will be— to take the matter from a healthy child, and from a well-formed vesicle, I consider it better than taking it *direct* from the cow, for the following reasons :—The cow-pox lymph, taken direct from the cow, produces much more violent symptoms than after it has passed through several persons; indeed, in some cases, it has produced effects as severe as cutting for the small-pox, besides, it has caused, in many cases, violent inflammation and even sloughing of the arm. There are also several kinds of *spurious* cow-pox to which the cow is subject, and which would be likely to be mistaken for the. *real* lymph. Again, if even the *genuine* matter were not taken from the cow *exactly* at the proper time, it would be deprived of its protecting power.

50. *At what age do you recommend an infant to be first vaccinated?*

When he is two months old, as the sooner he is protected the better. Moreover, the older he is the greater will be the difficulty in making him submit to the operation, and in preventing his arm from being rubbed, thus endangering the breaking of the vesicles, and thereby interfering with its effects. If small-pox be prevalent in the neighbourhood, he may, with perfect safety, be vaccinated at the month's end ; indeed if the small-pox be near at hand, he *must* be vaccinated, regardless of his age, and regardless of everything else, for small-pox spares neither the young nor the old, and if a new-born babe should unfortunately catch the disease, he will most likely die, as at his tender age he would not have strength to battle with such a formidable enemy. " A case, in the General Lying-in-Hospital, Lambeth, of small-pox occurred in a woman a few days after her admission, and the birth of her child. Her own child was vaccinated when only four days old, and all the other infants in the house varying from one day to a fortnight and more. All took the vaccination ; and the woman's own child, which suckled her and slept with her ; and all escaped the small pox."*

51. *Do you consider that the taking of matter from a child's arm weakens the effect of vaccination on the system?*

Certainly not, provided it has taken effect in more than

---

* Communicated by Sir Charles Locock to the Author.

one place. The arm is frequently much inflamed, and vaccinating other children from it abates the inflammation, and thus affords relief. *It is always well to leave one vesicle undisturbed.*

52. *If the infant have any " breaking out " upon the skin, ought that to be a reason for deferring the vaccination ?*

It should, as two skin diseases cannot well go on together ; hence the cow-pox might not take, or, if it did, might not have its proper effect in preventing small-pox. " It is essential that the vaccine bud or germ have a congenial soil, uncontaminated by another poison, which, like a weed, might choke its healthy growth."—*Dendy.* The moment the skin be free from the breaking-out, he must be vaccinated. A trifling skin affection, like red gum, unless it be severe, ought not, at the proper age to prevent ' vaccination. If small-pox be rife in the neighbourhood, the child *must* be vaccinated, regardless of *any* " breaking-out " on the skin.

53. *Does vaccination make a child poorly ?*

At about the fifth day after vaccination, and for three or four days, he is generally a little feverish ; the mouth is slightly hot, and he delights to have the nipple in his mouth. He does not rest so well at night ; he is rather cross and irritable ; and, sometimes, has a slight bowel-complaint. The arm, about the ninth or tenth day, is usually much inflamed—that is to say it is, for an inch or two or more around the vesicles, red, hot, swollen, and continues in this state for a day or two, at the end of which time the inflammation gradually subsides. It might be well to state that the above slight symptoms are desirable, as it proves that the vaccination has had a proper effect on his system, and that, consequently, he is more likely to be thoroughly protected from any risk of catching small-pox.

54. *Do you approve, either during or after vaccination, of giving medicine, more especially if he be a little feverish ?*

No, as it would be likely to work off some of its effects,

and thus would rob the cow-pox of its efficacy on the system. I do not like to interfere with vaccination *in any way whatever* (except, at the proper time, to take a little matter from the arm), but to allow the pock to have full power upon his constitution.

What do you give the medicine for ? If the matter that is put into the arm be healthy, what need is there of physic ! And if the matter be not of good quality, I am quite sure that no physic will make it so ! Look, therefore, at the case in whatever way you like, physic after vaccination is *not* necessary ; but, on the contrary, hurtful. If the vaccination produce slight feverish attack, it will, without the administration of a particle of medicine, subside in two or three days.

55. *Have you any directions to give respecting the arm* AFTER *vaccination ?*

The only precaution necessary is to take care that the arm be not rubbed ; otherwise the vesicles may be prematurely broken, and the efficacy of the vaccination may be lessened. The sleeve, in vaccination, ought to be large and soft, and should not be tied up. The tying up of a sleeve makes it hard, and is much more likely to rub the vesicles than if it were put on the usual way.

56. *If the arm,* AFTER *vaccination, be much inflamed, what ought to be done ?*

Smear frequently, by means of a feather or a camel's hair brush, a little cream on the inflamed part. This simple remedy will afford great comfort and relief.

57. *Have the goodness to describe the proper appearance, after the falling-off of the scab of the arm ?*

It might be well to remark, that the scabs ought always to be allowed to fall off of themselves. They must not, on any account, be picked or meddled with. With regard to the proper appearance of the arm, after the falling-off of the scab, " a perfect vaccine scar should be of small size, circular, and marked with radiations and indentations."—*Gregory.*

DENTITION.

58. *At what time does dentition commence ?*

The period at which it commences is uncertain. It may, as a rule, be said that a babe begins to cut his teeth at seven months old. Some have cut teeth at three months ; indeed, there are instances on record of infants having been born with teeth. King Richard the Third is said to have been an example. Shakspeare notices it thus :—

> " YORK.—Marry, they say my uncle grew so fast,
>   That he could gnaw a crust at two hours old
>   'Twas full two years ere I could get a tooth.
>   Grandam, this would have been a biting jest."

When a babe is born with teeth, they generally drop out. On the other hand, teething, in some children does not commence until they are a year and a half or two years old, and, in rare cases, not until they are three years old. There are cases recorded of adults who have never cut any teeth. An instance of the kind came under my own observation.

Dentition has been known to occur in old age. A case is recorded by M. Carre, in the *Gazette Médicale de Paris* (Sept. 15, 1860), of an old lady, aged eighty-five, who cut several teeth after attaining that age !

59. *What is the number of the* FIRST *set of teeth, and in what order do they generally appear ?*

The first or temporary set consists of twenty. The first set of teeth are usually cut in pairs. " I may say that nearly invariably the order is—1st, the lower front incissors [cutting teeth], then the upper front, then the *upper* two lateral incissors, and that not uncommonly a double tooth is cut before the two *lower* laterals ; but at all events the lower laterals come 7th and 8th, and, not 5th and 6th, as nearly all books on the subject testify."* Then the first grinders, in the lower jaw, afterwards the first upper grinders, then the lower corner-

---

* Sir Charles Locock in a *Letter* to the Author

pointed or canine teeth, after which the upper corner or eye-teeth, then the second grinders in the lower jaw, and lastly, the second grinders of the upper jaw. They do not, of course, always appear in this rotation. Nothing is more uncertain than the order of teething. A child seldom cuts his second grinders until after he is two years old. *He is usually, from the time they first appear, two years in cutting his first set of teeth.* As a rule, therefore, a child of two years old has sixteen, and one of two years and a half old, twenty teeth.

60. *If an infant be either feverish or irritable, or otherwise poorly, and if the gums be hot, swollen, and tender, are you an advocate for their being lanced?*

Certainly ; by doing so he will, in the generality of instances, be almost instantly relieved.

61. *But it has been stated that lancing the gums hardens them?*

This is a mistake—it has a contrary effect. It is a well-known fact, that a part which has been divided gives way much more readily than one which has not been cut. Again, the tooth is bound down by a tight membrane, which, if not released by lancing, frequently brings on convulsions. If the symptoms be urgent, it may be necessary from time to time to repeat the lancing. It would, of course, be the height of folly to lance the gums unless they be hot and swollen, and unless the tooth, or the teeth, be near at hand. It is not to be considered a panacea for every baby's ill, although, in those cases where the lancing of the gums is indicated, the beneficial effect is sometimes almost magical.

62. *How ought the lancing of a child's gums to be performed?*

The proper person, of course, to lance his gums is a medical man. But if, perchance, you should be miles away and be out of the reach of one, it would be well for you to know how the operation ought to be performed. Well, then, let him lie on the nurse's lap upon his back, and let the nurse take hold of his hands in order that he may not interfere with the operation.

Then, *if it be the upper gum* that requires lancing,
you ought to go to the head of the child, looking over,
as it were, and into his mouth, and should steady the
gum with the index finger of your left hand ; then, you
should take hold of the gum-lancet with your right
hand—holding as if it were a table-knife at dinner—and
cut firmly along the inflamed and swollen gum and
down to the tooth, until the edge of the gum-lancet
grates on the tooth.   Each incision ought to extend along
the ridge of the gum to about the extent of each
expected tooth.

If it be the lower gum that requires lancing, you must
go to the side of the child, and should steady the outside
of the jaw with the fingers of the left hand, and the gum
with the left thumb, and then you should perform the
operation as before directed.

Although the lancing of the gums, to make it intel-
ligible to a non-professional person, requires a long
description, it is, in point of fact, a simple affair, is soon
performed, and gives but little pain.

63. *If teething cause convulsions, what ought to be
done ?*

The first thing to be done (after sending for a medical
man) is to freely dash water upon the face, and to sponge
the head with cold water, and as soon as warm water can
be procured, to put him into a warm bath* of 98 degrees
Fahrenheit.   If a thermometer be not at hand,† you
must plunge your own elbow into the water : a comfortable heat for your elbow will be the proper heat for the
infant.   He must remain in the bath for a quarter of an
hour, or until the fit be at an end.   The body must,
after coming out of the bath, be wiped with warm and
dry and coarse towels ; he ought then to be placed in a
warm blanket.   The gums must be lanced, and cold
water should be applied to the head.   An enema, com-

---

* For the precautions to be used in putting a child into a warm
bath, see the answer to question on " Warm Baths."
† No family, where there are young children, should be without Fahrenheit's thermometer.

posed of table salt, of olive oil, and warm oatmeal gruel—
in the proportion of one table-spoonful of salt, of one
of oil, and a tea-cupful of gruel—ought then to be
administered, and should, until the bowels have been
well opened, be repeated every quarter of an hour; as
soon as he comes to himself a dose of aperient medicine
ought to be given.

It may be well, for the comfort of a mother, to state
that a child in convulsions is perfectly insensible to all
pain whatever; indeed, a return to consciousness speedily
puts convulsions to the rout.

64. *A nurse is in the habit of giving a child, who is
teething, either coral, or ivory, to bite : do you approve
of the plan ?*

I think it a bad practice to give him any hard,
unyielding substance, as it tends to harden the gums,
and, by so doing, causes the teeth to come through with
greater difficulty. I have found softer substances, such
as either a piece of wax taper, or an India-rubber ring,
or a piece of the best bridle leather, or a crust of bread,
of great service. If a piece of crust be given as a gum-
stick, he must, while biting it, be well watched, or by
accident he might loosen a large piece of it, which might
choke him. The pressure of any of these excites a more
rapid absorption of the gum, and thus causes the tooth
to come through more easily and quickly.

65. *Have you any objection to my baby, when he is
cutting his teeth, sucking his thumb ?*

Certainly not: the thumb is the best gum-stick in
the world :—it is convenient; it is handy (in every sense
of the word) : it is of the right size, and of the proper
consistence, neither too hard nor too soft; there is no
danger, as of some artificial gum-sticks, of its being
swallowed, and thus of its choking the child. The
sucking of the thumb causes the salivary glands to pour
out their contents, and thus not only to moisten the dry
mouth, but assist the digestion; the pressure of the
thumb eases, while the teeth are " breeding," the pain
and irritation of the gums, and helps, when the teeth are

sufficiently advanced, to bring them through the gums. Sucking of the thumb will often make a cross infant contended and happy, and will frequently induce a restless babe to fall into a sweet refreshing sleep. Truly may the thumb be called a baby's comfort. By all means, then, let your child suck his thumb whenever he likes, and as long as he chooses to do so.

There is a charming, bewitching little picture of a babe sucking his thumb in Kingsley's *Water Babies*, which I heartily commend to your favourable notice and study.

66. *But if an infant be allowed to suck his thumb, will it not be likely to become a habit, and stick to him for years—until, indeed, he become a big boy?*

After he have cut the whole of his first set of teeth, that is to say, when he is about two years and a half old, he might, if it be likely to become a habit, be readily cured by the following method, namely, by making a paste of aloes and water, and smearing it upon his thumb. One or two dressings will suffice as after just tasting the bitter aloes he will take a disgust to his former enjoyment, and the habit will at once be broken.

Many persons I know have an objection to children sucking their thumbs, as for instance,—

> " Perhaps it's as well to keep children from plums,
> And from pears in the season, and sucking their thumbs."[*]

My reply is,—

> P"rhaps 'tis as well to keep children from pears ;
> The pain they might cause, is oft follow'd by tears ;
> 'Tis certainly well to keep them from plums ;
> But certainly not from sucking their thumbs !
>> If a babe suck his thumb
>> 'Tis an ease to his gum ;
> A comfort ; a boon ; a calmer of grief ;
> A friend in his need—affording relief ;
> A solace ; a good ; a soother of pain ;
> A composer to sleep ; a charm ; and a gain.

---

* *Ingoldsby Legends.*

'Tis handy, at once, to his sweet mouth to glide;
When done with, drops·gently down by his side;
'Tis fix'd, like an anchor, while the babe sleeps,
And the mother, with joy, her still vigil keeps.

67. *A child who is teething dribbles, and thereby wets
his chest, which frequently causes him to catch cold;
what had better be done ?*

Have in readiness to put on several *flannel* dribbling
bibs, so that they may be changed as often as they
become wet; or, if he dribble *very much*, the oiled silk
dribbling-bibs, instead of the flannel ones, may be used,
and which may be procured at any baby-linen ware
house.

68. *Do you approve of giving a child, during teething,
much fruit ?*

No; unless it be a few ripe strawberries or raspberries,
or a roasted apple, or the juice of five or six grapes—
taking care that he does not swallow either the seeds or
the skin—or the insides of ripe gooseberries, or an
orange. Such fruits, if the bowels be in a costive state,
will be particularly useful.

All stone fruit, *raw* apples or pears, ought to be carefully avoided, as they not only disorder the stomach and
the bowels,—causing convulsions, gripings, &c.,—but
they have the effect of weakening the bowels, and thus
of engendering worms.

69. *Is a child, during teething, more subject to disease,
and, if so, to what complaints, and in what manner may
they be prevented?*

The teeth are a fruitful source of suffering and of
disease; and are, with truth, styled " our first and our
last plagues." Dentition is the most important period
of a child's life, and is the exciting cause of many
infantile diseases; during this period, therefore, he
requires constant and careful watching. When we consider how the teeth elongate and enlarge in his gums,
pressing on the nerves and on the surrounding parts, and
thus how frequently they produce pain, irritation, and
inflammation; when we further contemplate what

D

sympathy there is in the nervous system, and how susceptible the young are to pain, no surprise can be felt at the immense disturbance, and the consequent suffering and danger frequently experienced by children while cutting their *first* set of teeth. The complaints or the diseases induced by dentition are numberless, affecting almost every organ of the body,—the *brain*, occasioning convulsions, water on the brain, &c. ; the *lungs*, producing congestion, inflammation, cough, &c. ; the *stomach*, exciting sickness, flatulence, acidity, &c. ; the *bowels*, inducing griping, at one time costiveness, and at another time purging ; the *skin*, causing " breakings-out."

To prevent these diseases, means ought to be used to invigorate a child's constitution by plain, wholesome food, as recommended under the article of diet ; by exercise and fresh air ;* by allowing him, weather permitting, to be out of doors a great part of every day ; by lancing the gums when they get red, hot, and swollen ; by attention to the bowels, and if he suffer more than usual, by keeping them rather in a relaxed state by any simple aperient, such as either castor oil, or magnesia and rhubarb, &c. ; and, let me add, by attention to his temper : many children are made feverish and ill by petting and spoiling them. On this subject I cannot do better than refer you to an excellent little work entitled Abbot's *Mother of Home*, wherein the author proves the great importance of *early* training.

70. *Have the goodness to describe the symptoms and the treatment of Painful Dentition?*

Painful dentition may be divided into two forms—(1) the Mild ; and (2) the Severe. In the *mild* form the

---

* The young of animals seldom suffer from cutting their teeth —and what is the reason ? Because they live in the open air, and take plenty of exercise ; while children are frequently cooped up in close rooms, and are not allowed the free use of their limbs. The value of fresh air is well exemplified in the Registrar-General's Report for 1843 ; he says that in 1,000,000 deaths, from all diseases, 616 occur in the town from teething, while 120 only take place in the country from the same cause.

child is peevish and fretful, and puts his fingers, and everything within reach, to his mouth ; he likes to have his gums rubbed, and takes the breast with avidity; indeed it seems a greater comfort to him than ever. There is generally a considerable flow of saliva, and he has frequently a more loose state of bowels than is his wont.

Now, with regard to the more *severe* form of painful dentition :—The gums are red, swollen, and hot, and he cannot without expressing pain bear to have them touched, hence, if he be at the breast, he is constantly loosing the nipple. There is dryness of the mouth, although before there had been a great flow of saliva. He is feverish, restless, and starts in his sleep. His face is flushed. His head is heavy and hot. He is sometimes convulsed.* He is frequently violently griped and purged, and suffers severely from flatulence. He is predisposed to many and severe diseases.

The *treatment,* of the *mild* form, consists of friction of the gum with the finger, with a little "soothing syrup," as recommended by Sir Charles Locock ;† a tepid-bath of about 92 degrees Fahrenheit, every night at bed time; attention to diet and to bowels; fresh air and exercise. For the mild form, the above plan will usually be all that is required. If he dribble, and the bowels be relaxed, so much the better : the flow of saliva and the increased action of the bowels afford relief, and therefore must not be interfered with. In the *mild* form, lancing of the gums is not desirable. The gums ought not to be lanced, unless the teeth be near at hand, and unless the gums be red, hot, and swollen.

---

* See answer to Question 63.

† "Soothing syrup."—Some of them probably contain opiates, but a perfectly safe and useful one is a little Nitrate of Potass in syrup of Roses—one scruple to half an ounce."—*Communicated by Sir Charles Locock to the Author.* This "soothing syrup" is not intended to be given as a mixture : but to be used as an application to rub the gums with. It may be well to state, that it is a perfectly harmless remedy even if a little of it were swallowed by mistake.

In the *severe* form a medical man should be consulted early, as more energetic remedies will be demanded; that is to say, the gums will require to be freely lanced, warm baths to be used, and medicines to be given, to ward off mischief from the head, from the chest, and from the stomach.

If you are living in the town, and your baby suffers much from teething, take him into the country. It is wonderful what change of air to the country will often do, in relieving a child who is painfully cutting his teeth. The number of deaths in London, from teething, is frightful; it is in the country comparatively trifling.

71. *Should an infant be purged during teething, or indeed, during any other time, do you approve of either absorbent or astringent medicines to restrain it?*

Certainly not. I should look upon the relaxation as an effort of nature to relieve itself. A child is never purged without a cause; that cause, in the generality of instances, is the presence of either some undigested food, or acidity, or depraved motions, that want a vent.

The better plan is, in such a case, to give a dose of aperient medicine, such as either castor oil, or magnesia and rhubarb; and thus work it off. IF WE LOCK UP THE BOWELS, WE CONFINE THE ENEMY, AND THUS PRODUCE MISCHIEF.* If he be purged more than usual, attention should be paid to the diet—if it be absolutely necessary to give him artificial food while suckling—and care must be taken not to overload the stomach.

72. *A child is subject to a slight cough during dentition—called by nurses "tooth-cough"—which a parent would not consider of sufficient importance to consult a doctor about: pray tell me, is there any objection to a mother giving her child a small quantity either of syrup of white poppies, or of paregoric, to ease it?*

A cough is an effort of nature to bring up any secretion from the lining membrane of the lungs, or from the

---

* "I should put this in capitals, it is so important and so often mistaken."—*C. Locock,*

bronchial tubes, hence it ought not to be interfered with. I have known the administration of syrup of white poppies, or of paregoric, to stop the cough, and thereby to prevent the expulsion of the phlegm, and thus to produce either inflammation of the lungs, or bronchitis. Moreover, both paregoric and syrup of white poppies are, for a young child, dangerous medicines (unless administered by a judicious medical man), and *ought never to be given by a mother.*

In the month of April 1844, I was sent for, in great haste, to an infant, aged seventeen months, who was labouring under convulsions and extreme drowsiness, from the injudicious administration of paregoric, which had been given to him to ease a cough. By the prompt administration of an emetic he was saved.

73. *A child, who is teething, is subject to a " breaking-out," more especially behind the ears—which is most disfiguring, and frequently very annoying: what would you recommend ?*

I would apply no external application to cure it, as I should look upon it as an effort of the constitution to relieve itself ; and should expect, if the " breaking-out " were repelled, that either convulsions, or bronchitis, or inflammation of the lungs, or water on the brain, would be the consequence. The only plan I should adopt would be, to be more careful in his diet ; to give him less meat (if he be old enough to eat animal food), and to give him, once or twice a week, a few doses of mild aperient medicine ; and, if the irritation from the " breaking-out " be great, to bathe it, occasionally, either with a little warm milk and water, or with rose water.

EXERCISE.

74. *Do you recommend exercise in the open air for a baby ? and if so, how soon after birth ?*

I am a great advocate for his having exercise in the open air. " The infant in arms makes known its desire for fresh air, by restlessness ; it cries, for it cannot speak its wants ; is taken abroad and is quiet."

The age at which he ought to commence taking exercise will, of course, depend upon the season and upon the
weather. If it be summer, and the weather be fine, he
should be carried in the open air, a week or a fortnight
after birth; but if it be winter, he ought not on any
account to be taken out under the month, and not even
then, unless the weather be mild for the season, and it
be the middle of the day. At the end of two months he
should breathe the open air more frequently. And after
the expiration of three months, he ought to be carried
out *every day,* even if it be wet under foot, provided it
be fine above, and the wind be neither in an easterly nor
in a north-easterly direction: by doing so we shall make
him strong and hearty, and give the skin that mottled
appearance, which is so characteristic of health. He
must, of course, be well clothed.

I cannot help expressing my disapprobation of the
practice of smothering up an infant's face with a handkerchief, with a veil, or with any other covering, when he
is taken out into the air. If his face be so muffled up,
he may as well remain at home; as, under such circumstances, it is impossible for him to receive any benefit
from the invigorating effects of the fresh air.

75. *Can you devise any method to induce a babe himself to take exercise?*

He must be encouraged to use muscular exertion;
and, for this purpose, he ought to be frequently laid
either upon a rug, or carpet, or the floor: he will then
stretch his limbs and kick about with perfect glee. It
is a pretty sight, to see a little fellow kicking and spraw-
ling on the floor. He crows with delight and thoroughly
enjoys himself: it strengthens his back; it enables him
to stretch his limbs, and to use his muscles; and is one
of the best kinds of exercise a very young child can take.
While going through his performances, his diaper, if he
wear one, should be unfastened, in order that he might
go through his exercises untrammelled. By adopting the
above plan, the babe quietly enjoys himself—his brain
is not over excited by it: this is an important considera-

tion, for both mothers and nurses are apt to rouse, and excite very young children to their manifest detriment. A babe requires rest, and not excitement. How wrong it is, then, for either a mother or a nurse to be exciting and rousing a new-born babe. It is most injurious and weakening to his brain. In the early period of his existence his time ought to be almost entirely spent in sleeping and in sucking!

76. *Do you approve of tossing an infant much about?*
I have seen a child tossed nearly to the ceiling! Can anything be more cruel or absurd? Violent tossing of a young babe ought never to be allowed; it only frightens him, and has been known to bring on convulsions. He should be gently moved up and down (not tossed): such exercises causes a proper circulation of the blood, promotes digestion, and soothes to sleep. He must always be kept quiet immediately after taking the breast ; if he be tossed *directly* afterwards, it interferes with his digestion, and is likely to produce sickness.

SLEEP.

77. *Ought the infant's sleeping apartment to be kept warm?*
The lying-in room is generally kept too warm, its heat being, in many instances, more that of an oven than of a room. Such a place is most unhealthy, and is fraught with danger both to the mother and the baby. We are not, of course, to run into an opposite extreme, but are to keep the chamber at a moderate and comfortable temperature. The door ought occasionally to be left ajar, in order the more effectually to change the air and thus to make it more pure and sweet.

A new-born babe, then, ought to be kept comfortably warm, but not very warm. It is folly in the extreme to attempt to harden a very young child either by allowing him, in the winter time, to be in a bedroom without a fire, or by dipping him in *cold* water, or by keeping him with scant clothing on his bed. The temperature of a bedroom, in the winter time, should be, as nearly as possible, at 60° Fahr. Although the room should be comfortably warm, it ought from time to time to be properly ventilated. An unventilated room soon becomes foul, and, therefore, unhealthy. How many in this world, both children and adults, are " poisoned with their own breaths ! "

An infant should not be allowed to look at the glare either of a fire or of a lighted candle, as the glare tends to weaken the sight, and sometimes brings on an inflammation of the eyes. In speaking to, and in noticing a baby, you ought always to stand *before*, and not *behind* him, or it might make him squint.

78. *Ought a babe to lie alone from the first ?*

Certainly not : at first—say, for the first few months —he requires the warmth of another person's body, especially in the winter ; but care must be taken not to overlay him, as many infants, from carelessness in this particular, have lost their lives. After the first few months he had better lie alone, on a horse-hair mattress.

79. *Do you approve of rocking an infant to sleep ?*

I do not. If the rules of health be observed, he will sleep both soundly and sweetly without rocking ; if they be not, the rocking might cause him to fall into a feverish, disturbed slumber, but not into a refreshing, calm sleep. Besides, if you once take to that habit, he will not go to sleep without it.

80. *Then don't you approve of a rocking-chair, and of rockers to the cradle ?*

Certainly not: a rocking-chair, or rockers to the cradle, may be useful to a lazy nurse or mother, and may induce a child to sleep, but that restlessly, when he does not need sleep, or when he is wet and uncomfortable, and requires " changing ; " but will not cause him to have that sweet and gentle and exquisite slumber so characteristic of a baby who has no artificial appliances to make him sleep. No ! rockers are perfectly unnecessary, and the sooner they are banished the nursery the better will it be for the infant community. I do not know a more wearisome and monotonous sound than the everlasting rockings to and fro in some nurseries; they are often accompanied by a dolorous lullaby from the nurse, which adds much to the misery and depressing influence of the performance.

81. *While the infant is asleep, do you advise the head of the crib to be covered with a handkerchief, to shade his eyes from the light, and, if it be summer time, to keep off the flies?* *

If the head of the crib be covered, the babe cannot breathe freely; the air within the crib becomes contaminated, and thus the lungs cannot properly perform their functions. If his sleep is to be refreshing, he must breathe pure air. I do not even approve of a head to a crib. A child is frequently allowed to sleep on a bed with the curtains drawn completely close, as though it·were dangerous for a breath of air to blow upon him !* This practice is most injurious. An infant must have the full·benefit of the air of the room; indeed, the bed-room door· ought to be frequently left ajar, so that the air of the apartment may be changed; taking care, of course, not to expose him to a draught. If the·flies, while he is· asleep, annoy him, let a *net* veil be thrown over his face, as he can readily breathe through net, but not through a handkerchief.

82. *Have you any suggestions to offer as to the way a babe should be dressed when he is put down to sleep?*

Whenever he be·put down to sleep, be more than usually particular that his dress be loose in every part; be careful that there be neither strings nor bands, to cramp him. Let him, then, during repose, be more than ordinarily free and unrestrained—

> " If, whilst in cradled rest your infant sleeps,
> Your watchful eyes unceasing vigil keeps,
> Lest cramping bonds his pliant limbs constrain,
> And cause defects that manhood may retain."

---

* I have somewhere read that if a cage, containing a canary, be suspended at night within a bed where a person is sleeping, and the curtains be drawn closely around, that the bird will, in the morning, in all probability, be found dead !

83. *Is it a good sign for a young child to sleep much ?*

A babe who sleeps a great deal thrives much more than one who does not. I have known many children, who were born* small and delicate, but who slept the greatest part of their time, become strong and healthy. On the other hand, I have known those who were born large and strong, yet who slept but little, become weak and unhealthy.

The common practice of a nurse allowing a baby to sleep upon her lap is a bad one, and ought never to be countenanced. He sleeps cooler, more comfortably, and soundly in his crib.

The younger an infant is the more he generally sleeps, so that during the early months he is seldom awake, and then only to take the breast.

84. *How is it that much sleep causes a young child to thrive so well ?*

If there be pain in any part of the body, or if any of the functions be not properly performed, he sleeps but

---

* It may be interesting to a mother to know the average weight of new-born infants. There is a paper on the subject in the *Medical Circular* (April 10, 1861), and which has been abridged in *Braithwaite's Retrospect of Medicine* (July and December 1861). The following are extracts :—"Dr E. von Siebold presents a table of the weights of 3000 infants (1586 male and 1414 female), weighed immediately after birth. From this table (for which we have not space) it results that by far the greater number of the children, 2215 weighed between 6 and 8 lbs. From 5¾ to 6 lbs. the number rose from 99 to 268 ; and from 8 to 8¼ lbs. they fell from 226 to 67, and never rose again at any weight to 100. From 8¾ to 9¼ lbs. they sank from 61 to 8, rising, however, at 9½ lbs. to 21. Only six weighed 10 lbs., one 10¾ lbs., and two 11 lbs. The author has never but once met with a child weighing 11¾ lbs. The most frequent weight in the 3000 was 7 lbs., numbering 426. It is a remarkable fact, that until the weight of 7 lbs. the female infants exceeded the males in number, the latter thenceforward predominating. . . . . . . From these statements, and those of various other authors here quoted, the conclusion may be drawn that the normal weight of a mature new-born infant is not less than six nor more than 8 lbs., the average weight being 6½ or 7 lbs., the smaller number referring to female and the higher to male infants."

little. On the contrary, if there be exemption from pain, and if there be a due performance of all the functions, he sleeps a great deal; and thus the body becomes refreshed and invigorated.

85. *As much sleep is of such advantage, if an infant sleep but little, would you advise composing medicine to be given tb him?*

Certainly not. The practice of giving composing medicine to a young child cannot be too strongly reprobated. If he does not sleep enough, the mother ought to ascertain if the bowels be in a proper state, whether they be sufficiently opened, that the motions be of a good colour—namely, a bright yellow, inclining to orange colour—and free from slime or from bad smell. An occasional dose of rhubarb and magnesia is frequently the best composing medicine he can take.

86. *We often hear of coroner's inquests upon infants who have been found dead in bed—accidentally overlaid: what is usually the cause?*

Suffocation, produced either by ignorance, or by carelessness. From *ignorance* in mothers, in their not knowing the common laws of life, and the vital importance of free and unrestricted respiration, not only when babies are up and about, but when they are in bed and asleep. From *carelessness*, in their allowing young and thoughtless servants to have the charge of infants at night; more especially as young girls are usually heavy sleepers, and are thus too much overpowered with sleep to attend to their necessary duties.

A foolish mother sometimes goes to sleep while allowing her child to continue sucking. The unconscious babe, after a time, looses the nipple, and buries his head in the bed-clothes. She awakes in the morning, finding, to her horror, a corpse by her side, with his nose flattened, and a frothy fluid, tinged with blood, exuding from his lips! A mother ought, therefore, never to go to sleep until her child have finished sucking.

The following are a few rules to prevent an infant from being accidentally overlaid:—(1.) Let your baby while asleep have plenty of room in the bed. (2.) Do not allow him to be too near to you; or if he be unavoidably near you (from the small size of the bed), let his face be turned to the opposite side. (3.) Let him lie fairly either on his side, or on his back. (4.) Be careful to ascertain that his mouth be not covered with the bed-clothes; and, (5) Do not smother his face with clothes, as a plentiful supply of pure air is as necessary when he is awake, or even more so, than when he is asleep. (6.) Never let him lie low in the bed. (7.) Let there be *no* pillow near the one his head is resting on, lest he roll to it, and thus bury his head in it. Remember, a young child has neither the strength nor the sense to get out of danger; and, if he unfortunately either turn on his face, or bury his head in a pillow that is near, the chances are that he will be suffocated, more especially as these accidents usually occur at night, when the mother, or the nurse, is fast asleep. (8.) Never intrust him at night to a young and thoughtless servant.

THE BLADDER AND THE BOWELS OF AN INFANT.

87. *Have you any hints to offer respecting the bowels and the bladder of an infant during the first three months of his existence?*

A mother ought daily to satisfy herself as to the state of the bladder and the bowels of her child. She herself should inspect the motions, and see that they are of a proper colour (bright yellow, inclining to orange), and consistence (that of thick gruel), that they are neither slimy, nor curdled, nor green; if they should be either the one or the other, it is a proof that she herself has, in all probability, been imprudent in her diet, and that it will be necessary for the future that she be more careful both in what she eats and in what she drinks.

She ought, moreover, to satisfy herself that the urine does not smell strongly, that it does not stain the diapers, and that he makes a sufficient quantity.

A frequent cause of a child crying is, he is wet, and uncomfortable, and wants drying and changing, and the only way he has of informing his mother of the fact is by crying lustily, and thus telling her in most expressive language of her thoughtlessness and carelessness.

88. *How soon may an infant dispense with diapers ?*

A babe of three months and upwards, ought to be held out, at least, a dozen times during the twenty-four hours · if such a plan were adopted, diapers might at the end of three months be dispensed with—a great *desideratum*—and he would be inducted into clean habits—a blessing to himself, and a comfort to all around, and a great saving of dresses and of furniture. " Teach your children to be clean. A dirty child is the mother's disgrace."* Truer words were never written,—A DIRTY CHILD IS THE MOTHER'S DISGRACE.

AILMENTS, DISEASE, ETC.

89. *A new born babe frequently has a collection of mucus in the air passages, causing him to wheeze : is it a dangerous symptom ?*

No, not if it occur *immediately* after birth; as soon as the bowels have been opened, it generally leaves him, or even before, if he give a good cry, which as soon as he is born he usually does. If there be any mucus either within or about the mouth, impeding breathing, it must with a soft handkerchief be removed.

90. *Is it advisable, as soon as an infant is born, to give him medicine ?*

It is now proved that the giving of medicine to a babe *immediately* after birth is unnecessary, nay, that it is hurtful—that is, provided he be early put to the breast, as the mother's *first* milk is generally sufficient to open the bowels. Sir Charles Locock† makes the following sensible remarks on this subject :—" I used to limit any

---

* *Hints on Household Management,* By Mrs C. L. Balfour,
† In a *Letter* to the Author.

aperient to a new-born infant to those which had not the first milk; and who had wet-nurses, whose milk was, of course, some weeks old, but for many years I have never allowed any aperient at all to any new-born infant, and I am satisfied it is the safest and the wisest plan."

The advice of Sir Charles Locock—*to give no aperient to a new-born infant*—is most valuable, and ought to be strictly followed. By adopting his recommendation, much after misery might be averted. If a new-born babe's bowels be costive, rather than give him an aperient, try the effect of a little moist sugar, dissolved in a little water, that is to say, dissolve half a tea-spoonful of pure unadulterated *raw* sugar in a tea-spoonful of warm water and administer it to him ; if in four hours it should not operate, repeat the dose. Butter and raw sugar is a popular remedy, and is sometimes used by a nurse to open the bowels of a new-born babe, and where there is costiveness, answers the purpose exceedingly well, and is far superior to castor oil. Try by all means to do, if possible, without a particle of opening medicine. If you once begin to give aperients, you will have frequently to repeat them. Opening physic leads to opening physic, until at length his stomach and bowels will become a physic shop ! Let me, then, emphatically say, avoid, if possible, giving a new-born babe a drop or a grain of opening medicine. If from the first you refrain from giving an aperient, he seldom requires one afterwards. It is the *first* step, in this as in all other things, that is so important to take.

If a new-born babe have *not* for twelve hours made water, the medical man ought to be informed of it, in order that he may inquire into the matter, and apply the proper remedies. Be particular in attending to these directions, or evil consequences will inevitably ensue.

91. *Some persons say, that new-born female infants have milk in their bosoms, and that it is necessary to squeeze them, and apply plasters to disperse the milk.*

The idea of there being real milk in a baby's breast is doubtful, the squeezing of the bosom is barbarous, and the application of plasters is useless. "Without actually saying," says Sir Charles Locock, "there is milk secreted in the breasts of infants, there is undoubtedly not rarely considerable swelling of the breasts both in *female* and *male* infants, and on squeezing them a serous fluid oozes out. I agree with you that the nurses should never be allowed to squeeze them, but be ordered to leave them alone."*

92. *Have the goodness to mention the* SLIGHT *ailments which are not of sufficient importance to demand the assistance of a medical man ?*

I deem it well to make the distinction between *serious* and *slight* ailments; I am addressing a mother. With regard to serious ailments, I do not think myself justified, except in certain *urgent* cases, in instructing a parent to deal with them. It might be well to make a mother acquainted with the *symptoms*, but not with the *treatment*, in order that she might lose no time in calling in medical aid. This I hope to have the pleasure of doing in future Conversations.

Serious diseases, *with a few exceptions*, and which I will indicate in subsequent Conversations, ought never to be treated by a parent, not even in the *early* stages, for it is in the early stages that the most good can generally be done. It is utterly impossible for any one who is not trained to the medical profession to understand a *serious* disease in all its bearings, and thereby to treat it satisfactorily.

There are some exceptions to these remarks. It will be seen in future Conversations that Sir CHARLES LOCOCK considers that a mother ought to be made acquainted with the *treatment* of *some* of the more *serious* diseases, where delay in obtaining *immediate* medical assistance might be death. I bow to his superior judgment, and have supplied the deficiency in subsequent Conversations.

The ailments and the diseases of infants, such as may, in the absence of the doctor, be treated by a parent, are

---

* *Letter* to the Author.

the following :—Chafings, Convulsions, Costiveness, Flatulence, Gripings, Hiccup, Looseness of the Bowels (Diarrhœa), Dysentery, Nettle-rash, Red-gum, Stuffing of the Nose, Sickness, Thrush. In all these complaints I will tell you—*What to do,* and—*What* NOT *to do.*

93. *What are the causes and the treatment of Chafing?*

The want of water : inattention and want of cleanliness are the usual causes of chafing.

What to do.—The chafed parts ought to be well and thoroughly sponged with tepid *rain* water—allowing the water from a well-filled sponge to stream over them,—and, afterwards, they should be thoroughly, but tenderly, dried with a soft towel, and then be dusted, either with finely-powdered starch, made of wheaten flour, or with Violet Powder, or with finely-powdered Native Carbonate of Zinc, or they should be bathed with finely-powdered Fuller's-earth and tepid water

If, in a few days, the parts be not healed discontinue the above treatment, and use the following application : —Beat up well together the whites of two eggs, then add, drop by drop, two table-spoonfuls of brandy. When well mixed, put it into a bottle and cork it up. Before using it let the excoriated parts be gently bathed with luke-warm rain water, and, with a soft napkin, be tenderly dried ; then, by means of a camel's hair brush, apply the above liniment, having first shaken the bottle. But bear in mind, after all that can be said and done, *that there is nothing in these cases like water*—there is nothing like keeping the parts clean, and the only way of *thoroughly* effecting this object is *by putting him every morning* INTO *his tub.*

What NOT *to do.*—Do not apply white lead, as it is a poison. Do not be afraid of using *plenty* of water, as cleanliness is one of the most important items of the treatment.

94. *What are the causes of Convulsions of an infant?*

Stuffing him, in the early months of his existence, *with food,* the mother having plenty of breast-milk the while · the constant physicking of a child by his own mother; teething; hooping-cough, when attacking a very young baby.

I never knew a case of convulsions occur—say for the first four months—(except in very young infants labouring under hooping-cough), where children lived on the breast-milk alone, and where they were *not* frequently quacked by their mothers !

For the treatment of the convulsions from teething, see page 46.

What to do in a case of convulsions which has been caused by feeding an infant either with too much or with *artificial* food. Give him, every ten minutes, a teaspoonful of ipecacuanha wine, until free vomiting be excited, then put him into a warm bath (see Warm Baths); and when he comes out of it administer to him a teaspoonful of castor oil, and repeat it every four hours, until the bowels be well opened.

What NOT *to do.*—Do not, for at least a month after the fit, give him artificial food, but keep him entirely to the breast. Do not apply leeches to the head.

What to do in a case of convulsions from hooping cough.—There is nothing better than dashing cold water on the face, and immersing him in a warm bath of 98 degrees Fahr. If he be about his teeth, and they be plaguing him, let the gums be both freely and frequently lanced. Convulsions seldom occur in hooping-cough, unless the child be either very young or exceedingly delicate. Convulsions attending an attack of hooping-cough make it a *serious* complication, and requires the assiduous and skilful attention of a judicious medical man.

What NOT *to do* in such a case.—Do not apply leeches; the babe requires additional strength, and not to be robbed of it ; and do not attempt to treat the case yourself.

95. *What are the best remedies for the Costiveness of an infant ?*

I strongly object to the frequent administration of opening medicine, as the repetition of it increases the mischief to a tenfold degree.

What to do.—If a babe, after the first few months, were held out, and if, at regular intervals, he were put upon his chair, costiveness would not so much prevail. It is wonderful how soon the bowels, in the generality of cases, by this simple plan, may be brought into a regular state. Besides, it inducts an infant into clean habits. I know many careful mothers who have accustomed their children, after the first three months, to do without diapers altogether. It causes at first a little trouble, but that trouble is amply repaid by the good consequences that ensue ; among which must be named the dispensing with such encumbrances as diapers. Diapers frequently chafe, irritate, and gall the tender skin of a baby. But they cannot of course, at an early age be dispensed with, unless a mother have great judgment, sense, tact, and perseverance, to bring her little charge into the habit of having his bowels relieved and his bladder emptied every time he is either held out or put upon his chair.

Before giving an infant a particle of aperient medicine, try, if the bowels are costive, the effect of a little *raw* sugar and water, either half a tea-spoonful of raw sugar dissolved in a tea-spoonful or two of water, or give him, out of your fingers, half a tea-spoonful of raw sugar to eat. I mean by *raw* sugar, not the white, but the pure and unadulterated sugar, and which you can only procure from a respectable grocer. If you are wise, you will defer as long as you can giving an aperient. If you once begin, and continue it for a while, opening medicine becomes a dire necessity, and then woe betide the poor unfortunate child. Or, give a third of a tea-spoonful of honey, early in the morning, occasionally. Or, administer a warm water enema—a tablespoonful, or more, by means of a 2 oz. India Rubber Enema Bottle.

What NOT *to do.*—There are two preparations of mercury I wish to warn you against administering of your own accord, viz.—(1) Calomel, and a milder preparation called (2) Grey-powder (mercury with chalk). It is a common practice in this country to give calomel, on account of the readiness with which it can be administered it being small in quantity, and nearly tasteless. Grey · powder also, is, with many mothers, a favourite in the nursery. It is a medicine of immense power—either for good or for evil; in certain cases it is very valuable; but in others, and in the great majority, it is very detrimental. This practice, then, of a mother giving mercury, whether in the form either of calomel or of grey-powder, cannot be too strongly reprobated, as the frequent administration either of the one or of the other weakens the body, predisposes it to cold, and frequently excites king's-evil—a disease too common in this country. Calomel and grey-powder, then, ought never to be administered unless ordered by a medical man.

Syrup of buckthorn and jalap are also frequently given, but they are griping medicines for a baby, and ought to be banished from the nursery.

The frequent repetition of opening medicines, then, in any shape or form, very much interferes with digestion; they must, therefore, be given as seldom as possible.

Let me, at the risk of wearying you, again urge the importance of your avoiding, as much as possible, giving a babe purgative medicines. They irritate beyond measure the tender bowels of an infant, and only make him more costive afterwards; they interfere with his digestion, and are liable to give him cold. A mother who is always, of her own accord, quacking · her child with opening physic, is laying up for ·her unfortunate offspring a debilitated constitution—a miserable existence.

For further information on this important subject see the 3d edition of *Counsel to a Mother* (*being the companion volume of Advice to a Mother*), on the great importance of desisting from irritating, from injuring, and from making still more costive, the obstinate bowels of a costive child,—by the administration of opening medicine,—however gentle and well-selected the aperients might be. Oh, that the above advice could be heard, and be acted upon, through the length and the breadth of the land; how much misery and mischief would then be averted !

96. *Are there any means of preventing the Costiveness of an infant ?*

If greater care were paid to the rules of health, such as attention to diet, exercise in the open air, thorough ablution of the *whole* body—more especially when he is being washed—causing the water, from a large and well-filled sponge, to stream over the lower part of his bowels ; the regular habit of causing him, at stated periods, to be held out, whether he want or not, that he may solicit a stool. If all these rules were observed, costiveness would not so frequently prevail, and one of the miseries of the nursery would be done away with.

Some mothers are frequently dosing their poor unfortunate babes either with magnesia to cool them, or with castor oil to heal their bowels ! Oh, the folly of such practices ! The frequent repetition of magnesia, instead of cooling an infant, makes him feverish and irritable. The constant administration of castor oil, instead of healing the bowels, wounds them beyond measure. No ! it would be a blessed thing if a babe could be brought up without giving him a particle of opening medicine ; his bowels would then act naturally and well : but then, as I have just now remarked, a mother must be particular in attending to Nature's medicines—to fresh air, to exercise, to diet, to thorough ablution, &c. Until that time comes, poor unfortunate babies must be, occasionally, dosed with an aperient.

97. *What are the causes of, and remedies for, Flatulence ?*

Flatulence most frequently occurs in those infants who live on *artificial* food, especially if they be over-fed. I therefore beg to refer you to the precautions I have given, when speaking of the importance of keeping a child for the first five or six months *entirely* to the breast ; and, if that be not practicable, of the times of feeding, and of the *best* kinds of artificial food, and of those which are least likely to cause " wind."

What to do.—Notwithstanding these precautions, if the babe should still suffer, " One of the best and safest remedies for flatulence is Sal-volatile,—a tea-spoonful of a solution of one drachm to an ounce and a half of water."* Or, a little dill or aniseed may be added to the food—half a tea-spoonful of dill water. Or, take twelve drops of oil of dill, and two lumps of sugar ; rub them well in a mortar together ; then add, drop by drop, three table-spoonfuls of spring water ; let it be preserved in a bottle for use. A tea-spoonful of this, first shaking the vial, may be added to each quantity of food. Or, three tea-spoonfuls of bruised caraway-seeds may be boiled for ten minutes in a tea-cupful of water, and then strained. One or two tea-spoonfuls of the caraway-tea may be added to each quantity of his food, or a dose of rhubarb and magnesia may occasionally be given.

Opodeldoc, or warm olive oil, well rubbed, for a quarter of an hour at a time, by means of the warm hand, over the bowels, will frequently give relief. Turning the child over on his bowels, so that they may press on the nurses' lap, will often afford great comfort. A warm bath (where he is suffering severely) generally gives *immediate* ease in flatulence ; it acts as a fomentation to the bowels. But after all, a dose of mild aperient medicine, when the babe is suffering severely, is often the best remedy for " wind."

Remember, at all times, prevention, whenever it be —and how frequently it is—possible, is better than cure.

What NOT *to do.*—" Godfrey's Cordial," " Infants' Preservative," and " Dalby's Carminative," are sometimes given in flatulence ; but as most of these quack medicines contain, in one form or another, either opium or poppy, and as opium and poppy are both dangerous remedies for

---

* Sir Charles Locock, in a *Letter* to the Author. Since Sir Charles did me the honour of sending me, for publication, the above prescription for flatulence, a new "British Pharmacopœia" has been published, in which the sal-volatile is much increased in strength : it is therefore necessary to lessen the sal-volatile in the above prescription one-half—that is to say, a tea-spoonful of the solution of *half* a drachm to an ounce and a half of water.

children, ALL quack medicines must be banished the nursery.

Syrup of poppies is another remedy which is often given by a nurse to afford relief for flatulence ; but let me urge upon you the importance for banishing it from the nursery. It has (when given by unprofessional persons) caused the untimely end of thousands of children. The medical journals and the newspapers teem with cases of deaths from mothers incautiously giving syrup of poppies to ease pain and to procure sleep.

98. *What are the symptoms, the causes, and the treatment of " Gripings " of an infant ?*

The symptoms.—The child draws up his legs ; screams violently ; if put to the nipple to comfort him, he turns away from it and cries bitterly ; he strains, as though he were having a stool ; if he have a motion, it will be slimy, curdled, and perhaps green. If, in addition to the above symptoms, he pass a large quantity of watery fluid from his bowels, the case becomes one of *watery gripes*, and requires the immediate attention of a doctor.

The *causes* of "gripings" or "gripes" may proceed either from the infant or from the mother. If from the child, it is generally owing either to improper food or to over-feeding ; if from the mother, it may be traced to her having taken either greens, or pork, or tart beer, or sour porter, or pickles, or drastic purgatives.

What to do.—The *treatment*, of course, must depend upon the cause. If it arise from over-feeding, I would advise a dose of castor oil to be given, and warm fomentations to be applied to the bowels, and the mother, or the nurse, to be more careful for the future. If it proceed from improper food, a dose or two of magnesia and rhubarb in a little dill water, made palatable with simple syrup.* If it arise from a mother's imprudence in eating

---

* Take of—Powdered Turkey Rhubarb, half a scruple ;
    Carbonate of Magnesia, one scruple ;
    Simple Syrup, three drachms ;
    Dill Water, eight drachms ;
  Make a Mixture. One or two tea-spoonfuls (according to the

trash, or from her taking violent medicine, a warm bath: a warm bath, indeed, let the cause of "griping" be what it may, usually affords instant relief.

Another excellent remedy is the following:—Soak a piece of new flannel, folded into two or three thicknesses, in warm water; wring it tolerably dry, and apply as hot as the child can comfortably bear it to the bowels, then wrap him in a warm, dry blanket, and keep him, for at least half an hour, enveloped in it. Under the above treatment, he will generally soon fall into a sweet sleep, and awake quite refreshed.

What NOT *to do.*—Do not give opiates, astringents, chalk, or any quack medicine whatever.

If a child suffer from a mother's folly in her eating improper food, it will be cruel in the extreme for him a *second* time to be tormented from the same cause.

99. *What occasions Hiccup, and what is its treatment?*

Hiccup is of such a trifling nature as hardly to require interference. It may generally be traced to over-feeding. Should it be severe, four or five grains of calcined magnesia, with a little syrup and aniseed water, and attention to feeding, are all that will be necessary.

100. *Will you describe the symptoms of Infantile Diarrhœa?*

Infantile diarrhœa, or *cholera infantum,* is one of the most frequent and serious of infantile diseases, and carries off, during the year, more children than any other complaint whatever: a knowledge of the symptoms, therefore, is quite necessary for a mother to know, in order that she may, at the proper time, call in efficient medical aid.

It will be well, before describing the symptoms, to tell you how many motions a young infant ought to have a day, their colour, consistence, and smell. Well, then, he should have from three to six motions in the twenty-four hours; the colour ought to be a bright yellow, inclining to orange; the consistence should be that of thick

---

age of the child) to be taken every four hours, until relief be obtained—first shaking the bottle,

gruel ; indeed, his motion, if healthy, ought to be somewhat of the colour (but a little more orange-tinted) and
of the consistence of mustard made for the table ; it
should be nearly, if not quite, devoid of smell ; it ought
to have a faint and peculiar, but not a strong disagreeable
odour. If it have a strong and disagreeable smell, the
child is not well, and the case should be investigated,
more especially if there be either curds or lumps in the
motions ; these latter symptoms denote that the food
has not been properly digested.

Now, suppose a child should have a slight bowel complaint—that is to say, that he has six or eight motions
during the twenty-four hours,—and that the stools are
of a thinner consistence than what I have described,—
provided, at the same time, that he be not griped, that
he have no pain, and have not lost his desire for the
breast :—What ought to be done ? *Nothing.* A slight
looseness of the bowels should *never* be interfered with,
—it is often an effort of nature to relieve itself of some
vitiated motion that wanted a vent—or to act as a diversion, by relieving the irritation of the gums. Even if
he be not cutting his teeth, he may be "breeding" them
—that is to say, the teeth may be forming in his gums,
and may cause almost as much irritation as though he
were actually cutting them. Hence, you see the immense
good a slight " looseness of the bowels " may cause. I
think that I have now proved to you the danger of interfering in such a case, and that I have shown you the
folly and the mischief of at once giving astringents—such
as Godfrey's Cordial, Dalby's Carminative, &c.—to relieve
a *slight* relaxation.

A moderate " looseness of the bowels," then, is often
a safety-valve, and you may, with as much propriety,
close the safety-valve of a steam engine as stop a
moderate " looseness of the bowels ! "

Now, if the infant, instead of having from three to six
motions, should have more than double the latter
number ; if they be more watery ; if they become slimy
and green, or green in part and curdled ; if they should

have an unpleasant smell ; if he be sick, cross, restless, fidgety, and poorly ; if every time he have a motion he be griped and in pain, we should then say that he is labouring under Diarrhœa ; then, it will be necessary to give a little medicine, which I will indicate in a subsequent Conversation.

Should there be both blood and slime mixed with the stool, the case becomes more serious ; still, with proper care, relief can generally be quickly obtained. If the evacuations—instead of being stool—are merely blood and slime, and the child strain frequently and violently, endeavouring thus, but in vain, to relieve himself, crying at each effort, the case assumes the character of Dysentery.*

If there be a mixture of blood, slime, and stool from the bowels, the case would be called Dysenteric-diarrhœa. The latter case requires great skill and judgment on the part of a medical men, and great attention and implicit obedience from the mother and the nurse. I merely mention these diseases in order to warn you of their importance, and of the necessity of strictly attending to a doctor's orders.

101. *What are the causes of Diarrhœa—" Looseness of the bowels ? "*

Improper food ; overfeeding ; teething ; cold ; the mother's milk from various causes disagreeing, namely, from her being out of health, from her eating unsuitable food, from her taking improper and drastic purgatives, or from her suckling her child when she is pregnant. Of course, if any of these causes are in operation, they ought, if possible, to be remedied, or medicine to the babe will be of little avail.

102. *What is the treatment of Diarrhœa ?*

What to do.—If the case be *slight*, and has lasted two or three days (do not interfere by giving medicine at first), and if the cause, as it probably is, be some acidity or vitiated stool that wants a vent, and thus endeavours

---

* See Symptoms and Treatment of Dysentery.

to obtain one by purging, the best treatment is, to assist
nature by giving either a dose of castor oil, or a moderate
one of rhubarb and magnesia,* and thus to work off the
enemy.   After the enemy has been worked off, either by
the castor · oil, or by the magnesia and rhubarb, the
purging will, in all probability, cease ; but if the relaxation still continue, that is to say, for three or four days
—then, if medical advice cannot be procured, the following mixture should be given :—

> Take of—Aromatic Powder of Chalk and Opium, ten grains ;
>   Oil of Dill, five drops ;
>   Simple Syrup, three drachms ;
>   Water, nine drachms ;
> Make a Mixture.†   Half a tea-spoonful to be given to an infant
> of six months and under, and one tea-spoonful to a child above
> that age, every four hours—first shaking the bottle.

If the babe be at the breast, he ought, for a few days,
to be kept *entirely* to it.   The mother should be most
particular in her own diet.

What NOT *to do.*—The mother must neither take greens,
nor cabbage, nor raw fruit, nor pastry, nor beer ; indeed,
while the diarrhœa of her babe continues, she had better
abstain from wine, as well as from fermented liquors.
The child, if at the breast, ought not, while the diarrhœa
continues, to have any artificial food.   He must neither
be dosed with grey-powder (a favourite, but highly improper remedy, in these cases), nor with any quack
medicines, such as Dalby's Carminative or Godfrey's
Cordial.

103. *What are the symptoms of Dysentery ?*

Dysentery. frequently arises from a neglected diarrhœa.
It is more dangerous than diarrhœa, as it is of an
inflammatory character ; and as, unfortunately, it frequently attacks a delicate child, requires skilful handling ;
hence the care and experience required in treating a case
of dysentery.

---

* For a rhubarb and magnesia mixture prescription, see page
71 (*note*).

† Let the mixture be made by a chemist.

Well, then, what are the symptoms ? The infant, in all probability, has had an attack of diarrhœa—bowel complaint as it is called—for several days ; he having had a dozen or two of motions, many of them slimy and frothy, like "frog-spawn," during the twenty-four hours. Suddenly the character of the motion changes,—from being principally stool, it becomes almost entirely blood and mucus ; he is dreadfully griped, which causes him to strain violently, as though his inside would come away every time he has a motion,—screaming and twisting about, evidently being in the greatest pain, drawing his legs up to his belly and writhing in agony. Sickness and vomiting are always present, which still more robs him of his little remaining strength, and prevents the repair of his system. Now, look at his face ! It is the very picture of distress. Suppose he has been a plump, healthy little fellow, you will see his face, in a few days, become old-looking, care-worn, haggard, and pinched. Day and night the enemy tracks him (unless proper remedies be administered) ; no sleep, or if he sleep, he is, every few minutes, roused. It is heart-rending to have to attend a bad case of dysentery in a child,—the writhing, the screaming, the frequent vomiting, the pitiful look, the rapid wasting and exhaustion, make it more distressing to witness than almost any other disease a doctor attends.

104. *Can anything be done to relieve such a case ?*

Yes. A judicious medical man will do a great deal. But, suppose that you are not able to procure one, I will tell you *what to do* and *what* NOT *to do.*

What to do.—If the child be at the breast, keep him to it, and let him have nothing else, for dysentery is frequently caused by improper feeding. If your milk be not good, or it be scanty, *instantly* procure a healthy wet nurse. *Lose not a moment ;* for in dysentery, moments are precious. But, suppose that you have no milk, and that no wet-nurse can be procured : what then ? Feed him entirely on cow's milk—the milk of *one* healthy cow ; let the milk be unboiled, and be fresh from the cow. Give it in small quantities at a time, and frequently, so that it may be retained on the stomach. If a table-spoonful of the milk make him sick, give him a dessert-spoonful; if a dessert-spoonful cause sickness, let him only have a tea-spoonful at a time, and let it be repeated every quarter of an hour. But, remember, in such a case the breast milk—the breast milk alone—is incomparably superior to any other milk or to any other food whatever.

If he be a year old, and weaned, then feed him, as above recommended, on the cow's milk. If there be extreme exhaustion and debility, let fifteen drops of brandy be added to each table-spoonful of new milk, and let it be given every half hour.

Now with regard to medicine. I approach this part of the treatment with some degree of reluctance,—for dysentery is a case requiring opium—and opium I never like a mother of her own accord to administer. But suppose a medical man cannot be procured in time, the mother must then prescribe, or the child will die! *What then is to be done?* Sir Charles Locock considers " that, in severe dysentery, especially where there is sickness, there is no remedy equal to pure Calomel, in a full dose without opium."* Therefore, at the very *onset* of the disease, let from three to five grains (according to the age of the patient) of Calomel, mixed with an equal quantity of powdered white sugar, be put dry on the tongue. In three hours after let the following mixture be administered :—

Take of—Compound Powder of Ipecacuanha, five grains ;
Ipecacuanha Wine, one drachm;
Simple Syrup, three drachms ;
Cinnamon Water, nine drachms ;
To make a Mixture. A tea-spoonful to be given every three or four hours, first *well* shaking the bottle.

Supposing he cannot retain the mixture—the stomach rejecting it as soon as swallowed—what then? Give the

* Communicated by Sir Charles Locock to the Author.

opium, mixed with small doses of mercury with chalk and sugar, in the form of powder, and put one of the powders *dry* on the tongue, every three hours :—

> Take of—Powdered Opium, half a grain ;
> Mercury with chalk, nine grains ;
> Sugar of Milk, twenty-four grains ;
> Mix well in a mortar, and divide into twelve powders.

Now, suppose the dysentery has for several days persisted, and that, during that time, nothing but mucus and blood—that no real stool—has come from the bowels, then a combination of castor oil and opium* ought, instead of the medicine recommended above, to be given :—

> Take of—Mucilage of Gum Acacia, three drachms ;
> · Simple Syrup, three drachms ;
> Tincture of Opium, ten drops (*not* minims) ;
> Castor Oil, two drachms ;
> Cinnamon water, four drachms :
> Make a Mixture. A tea-spoonful to be taken every four hours, first *well* shaking the bottle.

A warm bath, at the commencement of the disease, is very efficacious ; but it must be given at the *commencement*. If he has had dysentery for a day or two, he will be too weak to have a warm bath ; then, instead of the bath, try the following :—Wrap him in a blanket, which has been previously wrung out of hot water; over which envelope him in a *dry* blanket. Keep him in this hot, damp blanket for half an hour ; then take him out, put on his night-gown and place him in bed, which has been, if it be winter time, previously warmed. The above " blanket treatment " will frequently give great relief, and will sometimes cause him to fall into a sweet sleep. A flannel bag, filled with hot powdered table salt, made hot in the oven, applied to the bowels, will afford much comfort.

---

* My friend, the late Dr Baly, who had made dysentery his particular study, considered the combination of opium and castor oil very valuable in dysentery.

What NOT *to do.*—Do not give aperients, unless it be, as before advised, the castor oil guarded with the opium; do not stuff him with artificial food; do not fail to send for a judicious and an experienced medical man; for, re- member, it requires a skilful doctor to treat a case of dysentery, more especially in a child.

105. *What are the symptoms, the causes, and the treatment of Nettle-rash ?*

Nettle-rash consists of several irregular, raised wheals, red at the base, and white on the summit, on different parts of the body; *but it seldom attacks the face.* It is not contagious, and it may occur at all ages and many times. It comes and goes, remaining only a short time in a place. It puts on very much the appearance of the child having been stung by nettles—hence its name. It produces great heat, itching, and irritation, sometimes to such a degree as to make him feverish, sick, and fretful. He is generally worse when he is warm in bed, or when the surface of his body is suddenly exposed to the air. Rubbing the skin, too, always aggravates the itching and the tingling, and brings out a fresh crop.

The *cause* of nettle-rash may commonly be traced to improper feeding; although, occasionally, it proceeds from teething.

What to do.—It is a complaint of no danger, and readily gives way to a mild aperient, and to attention to diet. There is nothing better to relieve the irritation of the skin than a warm bath. If it be a severe attack of nettle- rash, by all means call in a medical man.

What NOT *to do.*—Do not apply cold applications to his skin, and do not wash him (while the rash is out) in quite *cold* water. Do not allow him to be in a draught, but let him be in a well-ventilated room. If he be old enough to eat meat, keep it from him for a few days, and let him live on milk and farinaceous diet. Avoid strong purgatives, and calomel, and grey-powder.

106. *What are the symptoms and the treatment of Red-gum ?* ·

Red-gum, tooth-rash, red-gown, is usually owing to

irritation from teething; not always from the cutting but from the evolution—the "breeding," of the teeth. It is also sometimes owing to unhealthy stools irritating the bowels, and showing itself, by sympathy, on the skin. Red-gum consists of several small papulæ, or pimples, about the size of pins' heads, and may be known from measles—the only disease for which it is at all likely to be mistaken—by its being unattended by symptoms of cold, such as sneezing, running, and redness of the eyes, &c., and by the patches *not* assuming a crescentic—half-moon shape; red-gum, in short, may readily be known by the child's health being unaffected, unless, indeed, there be a great crop of pimples; then there will be slight feverishness.

What to do.—Little need be done. If there be a good deal of irritation, a mild aperient should be given. The child ought to be kept moderately, but not very warm.

What NOT *to do.*—Draughts of air, or cold should be carefully avoided; as, by sending the eruption suddenly in, either convulsions or disordered bowels might be produced. Do not dose him with grey-powder.

107. *How would you prevent " Stuffing of the nose " in a new-born babe?*

Rubbing a little tallow on the bridge of the nose is the old-fashioned remedy, and answers the purpose. It ought to be applied every evening just before putting him to bed. · If the " stuffing " be severe, dip a sponge in hot water, as hot as he can comfortably bear; ascertain that it be not too hot, by previously applying it to your own face, and then put it for a few minutes to the bridge of his nose. As soon as the hard mucus is within reach, it should be carefully removed.

108. *Do you consider sickness injurious to an infant?*

Many thriving babies are, after taking the breast, frequently sick; still we cannot look upon sickness otherwise than as an index of either a disordered or of an overloaded stomach. If the child be sick, and yet be thriving, it is a proof that he overloads his stomach. A mother, then, must not allow him to suck so much at a time. She should, until he retain all he takes, lessen the quantity of milk. If he be sick and does *not* thrive, the mother should notice if the milk he throws up has a sour smell; if it have, she must first of all look to her own health; she ought to ascertain if her own stomach be out of order; for if such be the case, it is impossible for her to make good milk. She should observe whether in the morning her own tongue be furred and dry; whether she have a disagreeable taste in her mouth, or pains at her stomach, or heart-burn, or flatulence. If she have all, or any of these symptoms, the mystery is explained why he is sick and does not thrive. She ought then to seek advice, and a medical man will soon put her stomach into good order; and, by so doing, will, at the same time, benefit her child.

But if the mother be in the enjoyment of good health, she must then look to the babe himself, and ascertain if he be cutting his teeth; if the gums require lancing; if the secretions from the bowels be proper both in quantity and in quality; and, if he have had *artificial* food—it being absolutely necessary to give such food—whether it agree with him.

What to do.—In the first place, if the gums be red, hot, and swollen, let them be lanced; in the second, if the secretion from the bowels be either unhealthy or scanty, give him a dose of aperient medicine, such as castor oil, or the following:—Take two or three grains of powdered Turkey rhubarb, three grains of pure carbonate of magnesia, and one grain of aromatic powder—Mix. The powder to be taken at bed-time, mixed in a tea-spoonful of sugar and water, and which should, if necessary, be repeated the following night. In the third place, if the food he be taking does not agree with him, change it (*vide* answer to question 33). Give it in smaller quantities at a time, and not so frequently; or what will be better still, if it be possible, keep him, for a while, entirely to the breast.

What NOT *to do.*—Do not let him overload his stomach either with breast milk, or *with artificial food.* Let the mother avoid, until his sickness be relieved, greens, cabbage, and all other green vegetables.

109. *What are the causes, the symptoms, the prevention, and the cure of Thrush ?*

The thrush is a frequent disease of an infant, and is often brought on either by stuffing or by giving him improper food. A child brought up *entirely*, for the first three or four months, on the breast, seldom suffers from this complaint. The thrush consists of several irregular, roundish, white specks on the lips, the tongue, the inside, and the angles of the mouth, giving the parts affected . the appearance of curds and whey having been smeared upon them. The mouth is hot and painful, and he is afraid to suck: the moment the nipple is put to his mouth he begins to cry. The thrush, sometimes, although but rarely, runs through the whole of the alimentary canal. It should be borne in mind that nearly every child, who is sucking, has his or her tongue white or "frosted," as it is sometimes called. The thrush may be mild or very severe.

. Now with regard to *What to do.*—As the thrush is generally owing to improper and to artificial feeding, *if the child be at the breast*, keep him, for a time, entirely to it. Do not let him be always sucking, as that will not only fret his mouth, but will likewise irritate and make sore the mother's nipple.

If he be not at the breast, but has been weaned, then keep him for a few days entirely to a milk diet—to the milk of one cow—either boiled, if it be hot weather, to keep it sweet; or unboiled, in cool weather—fresh as it comes from the cow, mixed with warm water.

The best medicine is the old-fashioned one of Borax, a combination of powdered lump-sugar and borax being a good one for the purpose : the powdered lump-sugar increases the efficacy, and the cleansing properties of the borax; it tends, moreover, to make it more palatable .—

Take of—Borax, half a drachm ;
Lump Sugar, two scruples;
To be well mixed together, and made into twelve powders.
One of the powders to be put dry on the tongue every four hours.

The best *local* remedy is Honey of Borax, which ought to be smeared frequently, by means of the finger, on the parts affected.

Thorough ventilation of the apartment must be observed; and great cleanliness of the vessels containing the milk should be insisted upon.

In a bad case of thrush, change of air to the country is most desirable; the effect is sometimes, in such cases, truly magical.

If the thrush be brought on either by too much or by improper food; in the first case of course, a mother must lessen the quantity; and, in the second, she should be more careful in her selection.

What NOT *to do.*—Do not use either a calf's teat or wash leather for the feeding-bottle; fortunately, since the invention of India-rubber teats, they are now nearly exploded; they were, in olden times, fruitful causes of thrush. Do not mind the trouble of ascertaining that the cooking-vessels connected with the baby's food are perfectly clean and sweet. Do not leave the purity and the goodness of the cow's milk (it being absolutely necessary to feed him on artificial food) to be judged either by the milk-man, or by the nurse, but taste and prove it yourself. Do not keep the milk in a warm place, but either in the dairy or in the cellar; and, if it be summer time, let the jug holding the milk be put in a crock containing lumps of ice. Do not use milk that has been milked longer than twelve hours, but if practicable, have it milked direct from the cow, and use it *immediately*—let it be really and truly fresh and genuine milk.

When the disease is *severe*, it may require more active treatment—such as a dose of calomel; *which medicine must never be given unless it be either under the direction of a medical man, or unless it be in an extreme case,— such as dysentery;* * therefore, the mother had better seek advice.

---

* See the Treatment of Dysentery.

In a *severe* case of thrush, where the complaint has been brought on by *artificial* feeding—the babe not having the advantage of the mother's milk—it is really surprising how rapidly a wet-nurse—if the case has not been too long deferred—will effect a cure, where all other means have been tried and have failed. The effect has been truly magical! In a severe case of thrush pure air and thorough ventilation are essential to recovery.

110. *Is anything to be learned from the cry of an infant?*

A babe can only express his wants and his necessities by a cry; he can only tell his aches and his pains by a cry; it is the only language of babyhood; it is the most ancient of all languages; it is the language known by our earliest progenitors; it is, if listened to aright, a very expressive language, although it is only but the language of a cry—

* "Soft i⎯ fancy, that nothing canst but cry."—*Shakspeare.*

There is, then, a language in the cry of an infant, which to a mother is the most interesting of all languages, and which a thoughtful medical man can well interpret. The cry of a child, to an experienced doctor, is, each and all, a distinct sound, and is as expressive as the notes of the gamut. The cry of passion, for instance, is a furious cry; the cry of sleepiness is a drowsy cry; the cry of grief is a sobbing cry; the cry of an infant when roused from sleep is a shrill cry; the cry of hunger is very characteristic,—it is unaccompanied with tears, and is a wailing cry; the cry of teething is a fretful cry; the cry of pain tells to the practised ear the part of pain; the cry of ear-ache is short, sharp, piercing, and decisive, the head being moved about from side to side, and the little hand being often put up to the affected side of the head; the cry of bowel-ache is also expressive,—the cry is not so piercing as from ear-ache, and is an interrupted, straining cry, accompanied with a drawing-up of the legs to the belly; the cry of bronchitis is a gruff and phlegmatic cry; the cry of inflammation of the lungs is more a moan than a cry; the cry of croup is hoarse, and rough, and ringing, and is so characteristic that it may truly be called "the croupy cry;" the cry of inflammation of the membranes of the brain is a piercing shriek —a danger signal—most painful to hear; the cry of a child recovering from a severe illness is a cross, and wayward, and tearful cry; he may truly be said to be in a quarrelsome mood; he bursts out, without rhyme or reason, into a passionate flood of tears—into "a tempest of tears:" tears are always, in a severe illness, to be looked upon as a good omen, as a sign of amendment, as—

> " The tears that heal and bless."—*H. Bonar.*

Tears, when a child is dangerously ill, are rarely, if ever, seen; a cry, at night, for light—a frequent cause of a babe crying—is a restless cry:—

> " An infant crying in the night;
> An infant crying for the light:
> And with no language but a cry."—*Tennyson.*

111. *If an infant be delicate, have you any objection to his having either veal or mutton broth, to strengthen him?*

Broths seldom agree with a babe at the breast. I have known them produce sickness, disorder the bowels, and create fever. I recommend you, therefore, not to make the attempt.

Although broth and beef-tea, when taken by the mouth, will seldom agree with an infant at the breast, yet, when used as an enema, and in small quantities, so that they may be retained, I have frequently found them to be of great benefit, they have in some instances appeared to have snatched delicate children from the brink of the grave.

112. *My baby's ankles are very weak: what do you advise to strengthen them?*

If his ankles be weak, let them every morning be bathed, after the completion of his morning's ablution, for five minutes each time, with bay-salt and water, a

small handful of bay-salt dissolved in a quart of rain
water (with the chill of the water off in the winter, and
of its proper temperature in the summer time) ; then let
them be dried ; after the drying, let the ankles be well
rubbed with the following liniment :—

> Take of—Oil of Rosemary, three drachms ;
> Liniment of Camphor, thirteen drachms :
> To make a Liniment.

Do not let him be put on his feet early ; but allow
him to crawl, and sprawl, and kick about the floor, until
his body and his ankles become strong.

Do not, on any account, without having competent
advice on the subject, use iron instruments, or mechanical supports of any kind : the ankles are generally, by
such artificial supports, made worse, in consequence of
the pressure causing a further dwindling away and en-
feebling of the ligaments of the ankles, already wasted
and weakened.

Let him wear shoes with straps over the insteps to
keep them on, and not boots : boots will only, by wasting
the ligaments, increase the weakness of the ankles.

113. *Sometimes there is a difficulty in restraining the
bleeding of leech bites. What is the best method ?*

The difficulty in these cases generally arises from the
improper method of performing it. For example—a
mother endeavours to stop the hæmorrhage by loading
the part with rag ; the more the bites discharge, the more
rag she applies. At the same time, the child probably
is in a room with a large fire, with two or three
candles, with the doors closed, and with perhaps
a dozen people in the apartment, whom the mother has,
in her fright, sent for. This practice is strongly
reprehensible.

If the bleeding cannot be stopped,—in the first place,
the fire must be extinguished, the door and windows
should be thrown open, and the room ought to be cleared
of persons, with the exception of one, or, at the most,
two ; and every rag should be removed. "Stopping of
leech bites.—The simplest and most certain way, till the

proper assistance is obtained, is the pressure of the finger, with nothing intervening. It *cannot* bleed through that."*

Many babies, by excessive loss of blood from leech bites, have lost their lives from a mother not knowing how to act, and also from the medical man either living at a distance, or not being at hand. Fortunately for the infantile community, leeches are now very seldom ordered by doctors.

114. *Supposing a baby to be poorly, have you any advice to give to his mother as to her own management?*

She must endeavour to calm her feelings or her milk will be disordered, and she will thus materially increase his illness. If he be labouring under any inflammatory disorder, she ought to refrain from the taking of beer, wine, and spirits, and from all stimulating food ; otherwise, she will feed his disease.

Before concluding the first part of my subject—the Management of Infancy—let me again urge upon you the importance—the paramount importance—if you wish your babe to be strong,and hearty,—of giving him as little opening physic as possible. The best physic for him is Nature's physic—fresh air, and exercise, and simplicity of living. A mother who is herself always drugging her child, can only do good to two persons— the doctor and the druggist !

If an infant from his birth be properly managed,—if he have an abundance of fresh air for his lungs,—if he have plenty of exercise for his muscles (by allowing him to kick and sprawl on the floor),—if he have a good swilling and sousing of water for his skin,—if, during the *early* months of his life, he have nothing but the mother's milk for his stomach,—he will require very little medicine—the less the better ! He does not want his stomach to be made into a doctor's shop ! The grand thing is not to take every opportunity of administering physic, but of using every means of with-

* Sir Charles Locock, in a *Letter* to the Author.

holding it! And if physic be necessary, not to doctor him yourself, unless it be in extreme and urgent cases (which in preceding and succeeding Conversations I either have or will indicate), but to employ an experienced medical man. A babe who is always, without rhyme or reason, being physicked, is sure to be puny, delicate, and unhealthy, and is ready at any moment to drop into an untimely grave!

I will maintain that a healthy child *never* requires drugging with opening physic, and that costiveness is brought on by bad management. Aperient medicines to a healthy child are so much poison! *Let me impress the above remarks on every mother's mind;* for it is a subject of vital importance. Never, then, give a purgative to a healthy child; for, if he be properly managed, he will never require one. If you once begin to give aperients, you will find a difficulty in discontinuing them. Finally, I will only say with *Punch,—* "Don't."

CONCLUDING REMARKS ON INFANCY.

115. In concluding the first part of our subject—Infancy—I beg to remark: there are four things essentially necessary to a babe's well-doing, namely, (1) plenty of water for his skin; (2) plenty of fresh genuine milk mixed with water for his stomach (of course, giving him ONLY his mother's milk during the first six, eight, or nine months of his existence); (3) plenty of pure air for his lungs; (4) plenty of sleep for his brain: these are the four grand essentials for an infant; without an abundance of one and all of them, perfect health is utterly impossible! Perfect health! the greatest earthly blessing, and more to be coveted than ought else beside! There is not a more charming sight in the universe than the beaming face of a perfectly healthy babe,—

" His are the joys of nature, his the smile,
The cherub smile, of innocence and health."—*Knox.*

PART II.

CHILDHOOD.

The child is father of the man.—WORDSWORTH.
Bairns are blessings.—SHAKSPEARE.
These are MY jewels!—CORNELIA.

ABLUTION.

116. *At twelve months old, do you still recommend a child to be* PUT IN HIS TUB *to be washed?*

Certainly I do, as I have previously recommended at page 6, in order that his skin may be well and thoroughly cleansed. If it be summer time, the water should be used cold; if it be winter, a dash of warm must be added, so that it may be of the temperature of new milk: but do not, on any account use *very warm* water. The head must be washed (but not dried) before he be placed in a tub; then, putting him in the tub (containing the necessary quantity of water, and washing him as previously recommended),* a large sponge should be filled with the water and squeezed over his head, so that the water may stream over the whole surface of his body. A jugful of water should, just before taking him out of his bath, be poured over and down his loins; all this ought rapidly to be done, and he must be quickly dried with soft towels, and then expeditiously dressed. For the washing of your child I would recommend you to use Castile soap in

---

* See Infancy—Ablution, page 6.

preference to any other; it is more pure, and less irritating, and hence does not injure the texture of the skin. Take care that the soap does not get into his eyes, or it might produce irritation and smarting.

117. *Some mothers object to a child's* STANDING *in the water.*

If the head be wetted before he be placed in the tub, and if he be washed as above directed, there can be no valid objection to it. He must not be allowed to remain in his tub more than five minutes.

118. *Does not washing the child's head, every morning, make him more liable to catch cold, and does it not tend to weaken his sight?*

It does neither the one nor the other; on the contrary, it prevents cold, and strengthens his sight; it cleanses his scalp, prevents scurf, and, by that means, causes a more beautiful head of hair. The head, after each washing, ought, with a soft brush, to be well brushed, but should not be combed. The brushing causes a healthy circulation of the scalp; but combing the hair makes the head scurfy, and pulls out the hair by the roots.

119. *If the head, notwithstanding the washing, be scurfy, what should be done?*

After the head has been well dried, let a little cocoa-nut oil be well rubbed, for five minutes each time, into the roots of the hair, and, afterwards, let the head be well brushed, but not combed. The fine-tooth comb will cause a greater accumulation of scurf, and will scratch and injure the scalp.

120. *Do you recommend a child to be washed* IN HIS TUB *every night and morning?*

No; once a day is quite sufficient; in the morning in preference to the evening; unless he be poorly, then, evening instead of morning; as, immediately after he has been washed and dried, he can be put to bed.

121. *Ought a child to be placed in his tub whilst he is in a state of perspiration?*

Not whilst he is perspiring *violently*, or the perspiration might be checked suddenly, and ill consequences would ensue; *nor ought he to be put in his tub when he is cold,* or his blood would be chilled, and would be sent from the skin to some internal vital part, and thus would be likely to light up inflammation—probably of the lungs. His skin, when he is placed in his bath, ought to be moderately and comfortably warm; neither too hot nor too cold.

122. *When the child is a year old, do you recommend cold or warm water to be used?*

If it be winter, a little warm water ought to be added, so as to raise the temperature to that of new milk. As the summer advances, less and less warm water is required, so that, at length, none is needed.

123. *If a child be delicate, do you recommend anything to be added to the water which may tend to brace and strengthen him?*

Either a handful of table-salt, or half a handful of bay-salt, or of Tidman's sea-salt, should be previously dissolved in a quart jug of *cold* water; then, just before taking the child out of his morning bath, let the above be poured over and down the back and loins of the child—holding the jug, while pouring its contents on the back, a foot distant from the child, in order that it might act as a kind of douche bath.

124. *Do you recommend the child, after he has been dried with the towel, to be rubbed with the hand?*

I do; as friction encourages the cutaneous circulation, and causes the skin to perform its functions properly, thus preventing the perspiration (which is one of the impurities of the body) from being sent inwardly either to the lungs or to other parts. The back, the chest, the bowels, and the limbs are the parts that ought to be well rubbed.

CLOTHING.

125. *Have you any remarks to make on the clothing of a child?*

Children, boys and girls, especially if they be delicate,

ought always to wear high dresses up to their necks. The exposure of the upper part of the chest (if the child be weakly) is dangerous. It is in the *upper* part of the lungs, in the region of the collar bones, that consumption first shows itself. The clothing of a child, more especially about the chest, should be large and full in every part, and be free from tight strings, so that the circulation of the blood may not be impeded, and that there may be plenty of room for the full development of the rapidly-growing body.

His frock, or tunic, ought to be of woollen material— warm, light, and porous, in order that the perspiration may rapidly evaporate. The practice of some mothers in allowing their children to wear tight bands round their waists, and tight clothes, is truly reprehensible.

Tight bands or *tight* belts around the waist of a child are very injurious to health; they crib in the chest, and thus interfere with the rising and the falling of the ribs—so essential to breathing. *Tight* hats ought never to be worn; by interfering with the circulation they cause headaches. Nature delights in freedom, and resents interference!

126. *What parts of the body in particular ought to be kept warm?*

The chest, the bowels, and the feet, should be kept comfortably warm. We must guard against an opposite extreme, and not keep them too hot. The head alone should be kept cool, on which account I do not approve either of night or of day caps.

127 *What are the best kinds of hat for a child?*

The best covering for the head, when he is out and about, is a loose-fitting straw hat, which will allow the perspiration to escape. It should have a broad rim, to screen the eyes. A sun-shade, that is to say, a sea-side hat—a hat made of cotton—with a wide brim to keep off the sun, is also an excellent hat for a child; it is very light, and allows a free escape of the perspiration. It can be bought, ready made, at a baby-linen warehouse

A knitted or crocheted woollen hat, with woollen rosettes to keep the ears warm, and which may be procured at any baby-linen warehouse, makes a nice and comfortable winter's hat for a child. It is also a good hat for him to wear while performing a long journey. The colour chosen is generally scarlet and white, which, in cold weather, gives it a warm and comfortable appearance.

It is an abominable practice to cover a child's head either with beaver or with felt, or with any thick impervious material. It is a well-ascertained fact, that both beaver and silk hats cause men to suffer from headache, and to lose their hair—the reason being, that the perspiration cannot possibly escape through them. Now, if the perspiration cannot escape, dangerous, or at all events injurious, consequences must ensue, as it is well known that the skin is a breathing apparatus, and that it will not with impunity bear interference.

Neither a child nor any one else should be permitted to be in the glare of the sun without his hat. If he be allowed, he is likely to have a sun-stroke, which might either at once kill him, or might make him an idiot for the remainder of his life; which latter would be the worse alternative of the two.

128. *Have you any remarks to make on keeping a child's hands and legs warm when in the winter time he is carried out?*

When a child either walks or is carried out in wintry weather, be sure and see that both his hands and legs are well protected from the cold. There is nothing for this purpose like woollen gloves, and woollen stockings ·coming up over the knees.

129. *Do you approve of a child wearing a flannel nightgown?*

He frequently throws the clothes off him, and has occasion to be taken up in the night, and if he have not a flannel gown on, is likely to catch cold; on which account I recommend it to be worn. The usual calico night-gown should be worn *under* it.

130. *Do you advise a child to be* LIGHTLY *clad, in order that he may be hardened thereby ?*

I should fear that such a plan, instead·of hardening, would be likely to produce a contrary effect. It is an ascertained fact that more children of the poor, who are thus lightly clad, die, than of those who are properly defended from the cold. Again, what holds good with a young plant is equally applicable to a young child ; and we all know that it is ridiculous to think of unnecessarily exposing a tender plant to harden it. If it were thus exposed, it would wither and die.

131. *If a child be delicate, if he have a cold body, or a languid circulation, or if he be predisposed to inflammation of the lungs, do you approve of his wearing flannel instead of linen shirts ?*

I do ; as flannel tends to keep the body at an equal temperature, thus obviating the effects of the sudden changes of the weather, and promotes by gentle friction the cutaneous circulation, thus warming the cold body, and giving an impetus to the languid circulation, and preventing an undue quantity of blood from being sent to the lungs, either to light up or to feed inflammation. *Fine* flannel, of course, ought to be worn, which should be changed as frequently as the usual shirts.

If a child have had an attack either of bronchitis or of inflammation of the lungs, or if he have just recovered from scarlet fever, by all means, if he have not previously worn flannel, *instantly* let him begin to do so, and let him, *next* to the skin, wear a flannel waistcoat. *This is important advice, and ought not to be disregarded.*

Scarlet flannel is now much used instead of *white* flannel ; and as scarlet flannel has a more comfortable appearance, and does not shrink so much in washing, it may be substituted for the white.

132. *Have you any remarks to make on the shoes and stockings of a child ? and on the right way of cutting the toe-nails ?*

He ought, during the winter, to wear lamb's wool stockings that will reach *above* the knees, and *thick* calico drawers that will reach a few inches *below* the knees ; as it is of the utmost importance to keep the lower extremities comfortably warm. It is really painful to see how many mothers expose the bare legs of their little ones to the frosty air, even in the depths of winter.

Be sure and see that the boots and shoes of your child be sound and whole ; for if they be not so, they will let in the damp, and if the damp, disease and perhaps death. " If the poor would take better care of their children's feet half the infantile mortality would disappear. It only costs twopence to put a piece of thick felt or cork into the bottom of a boot or shoe, and the difference is often between that and a doctor's bill, with, perhaps, the undertaker's besides."—*Daily Telegraph.*

Garters ought not to be worn, as they impede the circulation, waste the muscles, and interfere with walking. The stocking may be secured in its place by means of a loop and tape, which should be fastened to a part of the dress.

Let me urge upon you the importance of not allowing your child to wear *tight* shoes ; they cripple the feet, causing the joints of the toes, which ought to have free play, and which should assist in walking, to be, in a manner, useless ; they produce corns and bunions, and interfere with the proper circulation of the foot. A shoe ought to be made according to the shape of the foot—rights and lefts are therefore desirable. The toe-part of the shoe must be made broad, so as to allow plenty of room for the toes to expand, and that one toe cannot overlap another. Be sure, then, that there be no pinching and no pressure. In the article of shoes you ought to be particular and liberal ; pay attention to having nicely fitting ones, and let them be made of soft leather, and throw them on one side the moment they are too small. It is poor economy, indeed, because a pair of shoes be not worn out, to run the risk of incurring the above evil consequences.

Shoes are far preferable to boots : boots weaken instead of strengthen the ankle. The ankle and instep require free play, and ought not to be hampered by boots. Moreover, boots, by undue pressure, decidedly waste away the ligaments of the ankle. Boots act on the ankles in a similar way that stays do on the waist—they do mischief by pressure. Boots waste away the ligaments of the ankle; stays waste away the muscles of the back and chest; and thus, in both cases, do irreparable mischief.

A shoe for a child ought to be made with a narrow strap over the instep, and with button and button-hole; if it be not made in this way, the shoe will not keep on the foot.

It is a grievous state of things, that in the nineteenth century there are but few shoemakers who know how to make a shoe! The shoe is made not to fit a real foot, but a fashionable imaginary one! The poor unfortunate toes are in consequence screwed up as in a vice!

Let me strongly urge you to be particular that the sock, or stocking, fits nicely—that it is neither too small nor too large; if it be too small, it binds up the toes unmercifully, and makes one toe to ride over the other, and thus renders the toes perfectly useless in walking; if it be too large, it is necessary to lap a portion of the sock, or stocking, either under or over the toes, which thus presses unduly upon them, and gives pain and annoyance. It should be borne in mind, that if the toes have full play, they, as it were, grasp the ground, and greatly assist in locomotion—which, of course, if they are cramped up, they cannot possibly do. Be careful, too, that the toe-part of the sock, or stocking, be not pointed; let it be made square in order to give room to the toes. "At this helpless period of life, the delicately feeble, outspreading toes are wedged into a narrow-toed stocking, often so short as to double in the toes, diminishing the length of the rapidly growing foot! It is next, perhaps, tightly laced into a boot of less interior dimensions than itself; when the poor little creature is left to sprawl about with a limping, stumping gait. thus learning to walk as it best can, under circumstances the most cruel and torturing imaginable."*

It is impossible for either a stocking, or a shoe, to fit nicely unless the toe-nails be kept in proper order. Now, in cutting the toe-nails, there is, as in everything else, a right and a wrong way. The *right* way of cutting a toe-nail is to cut it straight—in a straight line. The *wrong* way is to cut the corners of the nail—to round the nail as it is called. This cutting the corners of the nails often makes work for the surgeon, as I myself can testify; it frequently produces "growing-in" of the nail, which sometimes necessitates the removal of either the nail, or a portion of it.

133. *At what time of the year should a child leave off his winter clothing ?*

A mother ought not to leave off her children's winter clothing until the spring be far advanced : it is far better to be on the safe side, and to allow the winter clothes to be worn until the end of May. The old adage is very good, and should be borne in mind :—

> " Button to chin
> Till May be in ;
> Ne'er cast a clout
> Till May be out."

134. *Have you any general remarks to make on the present fashion of dressing children ?*

The present fashion is absurd. Children are frequently dressed like mountebanks, with feathers and furbelows and finery ; the boys go bare-legged ; the little girls are dressed like women, with their stuck-out petticoats, crinolines, and low dresses ! Their poor little waists are drawn in tight, so that they can scarcely breathe ; their dresses are very low and short, the consequence is, that a great part of the chest is exposed to

---

* *The Foot and its Covering*, second edition. By James Dowie. London : 1872. I beg to call a mother's especial attention to this valuable little book : it is written by an earnest intelligent man, by one who has studied the subject in all its bearings, and by one who is himself a shoemaker.

our variable climate; their legs are bare down to their thin socks, or if they be clothed, they are only covered with gossamer drawers; while their feet are encased in tight shoes of paper thickness! Dress! dress! dress! is made with them, at a tender age, and when first impressions are the strongest, a most important consideration. They are thus rendered vain and frivolous, and are taught to consider dress "as the one thing needful." And if they live to be women—which the present fashion is likely frequently to prevent—what are they? Silly, simpering, delicate, lack-a-daisical nonentities; dress being their amusement, their occupation, their conversation, their everything, their thoughts by day and their dreams by night! Truly they are melancholy objects to behold! Let children be dressed as children, not as men and women. Let them be taught that dress is quite a secondary consideration. Let health, and not fashion, be the first, and we shall then have, with God's blessing, blooming children, who will, in time, be the pride and strength of dear old England!

DIET.

135. *At* TWELVE *months old, have you any objection to a child having any other food besides that you mentioned in answer to the* 34th *question?*

There is no objection to his *occasionally* having, for dinner, either a mealy, *mashed* potato and gravy, or a few crumbs of bread and gravy. Rice-pudding or batter-pudding may, for a change, be given; but remember, the food recommended in a former Conversation is what, until he be eighteen months old, must be principally taken. During the early months of infancy—say, for the first six or seven—if artificial food be given at all, it should be administered by means of a feeding-bottle. After that time, either a spoon, or a nursing boat, will be preferable. The food, as he becomes older, ought to be made more solid.

136. *At* EIGHTEEN *months old, have you any objection to a child having meat ?*

He ought not to have meat until he have several teeth to chew it with. If he has most of his teeth—which he very likely at this age will have—there is no objection to his taking a small slice either of mutton, or occasionally of roast beef, which should be well cut into very small pieces, and mixed with a mealy *mashed* potato, and a few crumbs of bread and gravy ; either *every* day, if he be delicate, or every *other* day, if he be a gross or a fast-feeding child. It may be well, in the generality of cases, for the first few months to give him meat *every other* day, and either potato or gravy, or rice or suet-pudding or batter-pudding on the alternate days ; indeed, I think so highly of rice, of suet, and of batter-puddings, and of other farinaceous puddings, that I should advise you to let him have either the one or the other even on those days that he has meat—giving it him *after* his meat. But remember, if he have meat *and* pudding, the meat ought to be given sparingly. If he be gorged with food, it makes him irritable, cross, and stupid ; at one time, clogging up his bowels, and producing constipation ; at another, dis-ordering his liver, and causing either clay-coloured stools—denoting a *deficiency* of bile, or dark and offensive motions—telling of *vitiated* bile ; while, in a third case, cramming him with food might bring on convulsions.

137. *As you are so partial to puddings for a child, which do you consider the best for him ?*

He ought, every day, to have a pudding for his dinner—either rice, arrow-root, sago, tapioca, suet-pudding, batter-pudding, or Yorkshire-pudding, mixed with crumbs of bread and gravy—free from grease. A well boiled suet-pudding, with plenty of suet in it, is one of the best puddings he can have ; it is, in point of fact, meat and farinaceous food combined, and is equal to, and will oftentimes prevent the giving of, cod-liver oil ; before cod-liver oil came into vogue, suet boiled in milk was *the* remedy for a delicate child. He may, occasionally, have fruit-pudding, provided the pastry be both plain and light.

The objection to fruit pies and puddings is, that the pastry is often too rich for the delicate stomach of a child; there is no objection, certainly not, to the fruit— cooked fruit being, for a child, most wholesome; if, therefore, fruit puddings and pies be eaten, the pastry part ought to be quite plain. There is, in "Murray's Modern Cookery Book," an excellent suggestion, which I will take the liberty of quoting, and of strongly urging my fair reader to carry into practice :—" *To prepare fruit for children, a far more wholesome way than in pies and puddings*, is to put apples sliced, or plums, currants, gooseberries, &c., into a stone jar; and sprinkle among them as much Lisbon sugar as necessary. Set the jar on an oven or on a hearth, with a tea-cupful of water to prevent the fruit from burning; or put the jar into a saucepan of water, till its contents be perfectly done. Slices of bread or some rice may be put into the jar, to eat with the fruit."

Jam—such as strawberry, raspberry, gooseberry—*is most wholesome for a child*, and ought occasionally to be given, in lieu of sugar, with the rice, with the batter, and with the other puddings. Marmalade, too, is very wholesome.

Puddings ought to be given *after* and not *before* his meat and vegetables; if you give him pudding before his meat, he might refuse to eat meat altogether. By adopting the plan of giving puddings *every* day, your child will require *less* animal food; *much* meat is injurious to a young child. But do not run into an opposite extreme : a *little* meat ought, every day, to be given, *provided he has cut the whole of his first set of teeth;* until then, meat every *other* day will be often enough.

138. *As soon as a child has cut the whole of his first set of teeth, what ought to be his diet? What should be his breakfast?*

He can, then, have nothing better, where it agrees, than

scalding hot new milk poured on sliced bread, with a slice or two of bread and butter to eat with it. Butter, in moderation, is nourishing, fattening, and wholesome. Moreover, butter tends to keep the bowels regular. These facts should be borne in mind, as some mothers foolishly keep their children from butter, declaring it to be too rich for their children's stomachs! New milk should be used in preference either to cream or to skim-milk. Cream, as a rule, is too rich for the delicate stomach of a child, and skim-milk is too poor when robbed of the butter which the cream contains. But give cream and water, where new milk (as is *occasionally* the case) does not agree; but never give skim-milk. *Skim*-milk (among other evils) produces costiveness, and necessitates the frequent administration of aperients. Cream, on the other hand, regulates and tends to open the bowels.

Although I am not, as a rule, so partial to cream as I am to good genuine fresh milk, yet I have found, in cases of great debility, more especially where a child is much exhausted by some inflammatory disease, such as inflammation of the lungs, the following food most serviceable :—Beat up, by means of a fork, the yolk of an egg, then mix, little by little, half a tea-cupful of very weak *black* tea, sweeten with one lump of sugar, and add a table-spoonful of cream. Let the above, by tea-spoon-fuls at a time be frequently given. The above food is only to be administered until the exhaustion be removed, and is not to supersede the milk diet, which must, at stated periods, be given, as I have recommended in answers to previous and subsequent questions.

When a child has costive bowels, there is nothing better for his breakfast than well-made and well-boiled oatmeal stir-about, which ought to be eaten with milk fresh from the cow. Scotch children scarcely take anything else, and a finer race is not in existence ; and, as for physic, many of them do not even know either the taste or the smell of it ! You will find Robinson's Pure Scotch Oat-meal (sold in packets) to be very pure, and sweet, and good. Stir-about is truly said to be—

" The halesome parritch, chief of Scotia's food."—*Burns.*

Cadbury's Cocoa Essence, made with equal parts of boiling water and fresh milk, slightly sweetened with lump sugar, is an admirable food for a delicate child. Bread and butter should be eaten with it.

139. *Have you any remarks to make on cow's milk as an article of food ?*

Cow's milk is a valuable, indeed, an indispensable article of diet, for the young ; it is most nourishing, wholesome, and digestible. The finest and the healthiest children are those who, for the first four or five years of their lives, are fed *principally* upon it. Milk ought then to be their staple food. No child, as a rule, can live, or, if he live, can be healthy, unless milk be the staple article of his diet. There is no substitute for milk. To prove the fattening and strengthening qualities of milk, look only at a young calf who lives on milk, and on milk alone ! He is a Samson in strength, and is " as fat as butter ;" and all young things if they are in health are fat !

Milk, then, contains every ingredient to build up the body, which is more than can be said of any other known substance besides. A child may live entirely, and grow, and become both healthy and strong, on milk, and on milk alone, as it contains every constituent of the human body. A child cannot "live by bread alone," but he might on milk alone ! Milk is animal and vegetable—it is meat and bread—it is food and drink—it is a fluid, but as soon as it reaches the stomach it becomes a solid*

---

* How is milk in the making of cheese, converted into curds ? By rennet. What is rennet ? The juice of a calf's maw or stomach. The moment the milk enters the human maw or stomach, the juice of the stomach converts it into curds—into solid food, just as readily as when it enters a calf's maw or stomach, and much more readily than by rennet, as the *fresh* juice is stronger than the *stale.* An ignorant mother often complains that because, when her child is sick, the milk curdles, that it is a proof that it does not agree with him ! If, at those times, it did *not* curdle, it would, indeed, prove that his stomach was in a wretchedly weak state ; she would then have abundant cause to be anxious.

—solid food; it is the most important and valuable
article of diet for a child in existence. It is a glorious
food for the young, and must never, on any account
whatever, in any case be dispensed with. "Considering
that milk contains in itself most of the constituents of a
perfect diet, and is capable of maintaining life in infancy
without the aid of any other substance, it is marvellous
that the consumption of it is practically limited to so
small a class; and not only so, but that in sick-rooms,
where the patient is surrounded with every luxury,
arrow-root, and other compounds containing much less
nutriment, should so often be preferred to it."—*The
Times.*

Do not let me be misunderstood. I do not mean to
say, but that the mixing of farinaceous food—such as
Lemann's Biscuit Powder, Robb's Biscuit, Hard's
Farinaceous Food, Brown and Polson's Corn Flour, and
the like, with the milk, is an improvement, in some
cases—a great improvement; but still I maintain that a
child might live and thrive, and that for a lengthened
period, on milk—and on milk alone!

A dog will live and fatten for six weeks on milk
alone; while he will starve and die in a shorter period
on strong beef-tea alone!

It is a grievous sin for a milkman to adulterate milk.
How many a poor infant has fallen a victim to that
crime!—for crime it may be truly called.

· It is folly in the extreme for a mother to bate a milk-
man down in the price of his milk; if she does, the
milk is sure to be either of inferior quality, or
adulterated, or diluted with water; and woe betide the
poor unfortunate child if it be either the one or the
other! The only way to insure good milk is, to go to a
respectable cow-keeper, and let him be made to
thoroughly understand the importance of your child
having *genuine* milk, and that you are then willing to
pay a fair remunerative price for it. Rest assured, that
if you have to pay one penny or even twopence a quart
more for *genuine* milk, it is one of the best investments

that you ever have made, or that you are ever likely to make in this world ! Cheap and inferior milk might well be called cheap and nasty ; for inferior or adulterated milk is the very essence, the conglomeration of nastiness ; and, moreover, is very poisonous to a child's stomach. One and the principal reason why so many children are rickety and scrofulous, is the horrid stuff called milk that is usually given to them. It is a crying evil, and demands a thorough investigation and reformation, and the individual interference of every parent. Limited Liability Companies are the order of the day ; it would really be not a bad speculation if one were formed in every large town, in order to insure good, genuine, and undiluted milk.

Young children, as a rule, are allowed to eat too much meat. It is a mistaken notion of a mother that they require so much animal food. If more milk were given and less meat, they would be healthier, and would not be so predisposed to disease, especially to diseases of debility, and to skin-disease.

I should strongly recommend you, then, to be extravagant in your milk score. Each child ought, in the twenty-four hours, to take at least a quart of good, fresh, new milk. It should, of course, be given in various ways,—as bread and milk, rice-puddings, milk and differents kinds of farinaceous food, stir-about, plain milk, cold milk, hot milk, any way, and every way, that will please his palate, and that will induce him to take an abundant supply of it. The " advice " I have just given you is of paramount importance, and demands your most earnest attention. There would be very few rickety children in the world if my " counsel " were followed out to the very letter.

140. *But suppose my child will not take milk, he having an aversion to it, what ought then to be done ?*

Boil the milk, and sweeten it to suit his palate. After he has been accustomed to it for a while, he will then, probably, like milk. Gradually reduce the sugar, until at length it be dispensed with. A child will often

take milk this way, whereas he will not otherwise touch
it.

If a child will not drink milk, he *must* eat meat; it is
absolutely necessary that he should have either the one
or the other ; and, if he have cut nearly all his teeth, he
ought to have both meat and milk—the former in
moderation, the latter in abundance.

141. *Supposing milk should not agree with my child,
what must then be done ?*

Milk, either boiled or unboiled, almost always agrees
with a child. If it does not, it must be looked upon as
the exception, and not as the rule. I would, in such a
case, advise one-eighth of lime water to be added to
seven-eighths of new milk—that is to say, two table-
spoonfuls of lime water should be mixed with half a pint
of new milk.

142. *Can you tell me of a way to prevent milk, in hot
weather, from turning sour ?*

Let the jug of milk be put into a crock, containing
ice—Wenham Lake is the best—either in the dairy or in
the cellar. The ice may at any time be procured of a
respectable fishmonger, and should be kept, wrapped
either in flannel or in blanket, in a cool place, until it be
wanted.

143. *Can you tell me why the children of the rich
suffer so much more from costiveness than do the children
of the poor ?*

The principal reason is that the children of the rich
drink milk without water, while the children of the poor
drink water without, or with very little, milk—milk
being binding, and water opening to the bowels. Be
sure then, and bear in mind, *as this is most important
advice,* to see that water is mixed with *all* the milk that
is given to your child. The combination of milk and
water for a child is a glorious compound—strengthening,
fattening, refreshing, and regulating to the bowels, and
thus doing away with that disgraceful proceeding so
common in nurseries, of everlastingly physicking, irritating and irreparably injuring the tender bowels of a child.

My opinion is, that aperients, as a rule, are quite unnecessary, and should only be given in severe illness, and under the direction of a judicious medical man. How much misery, and injury, might be averted if milk were always given to a child in combination with water !

Aperients, by repetition, unlike water, increase the mischief tenfold, and cork them up most effectually ; so that the bowels, in time, will not act without them !

A mother before she gives an aperient to her child should ponder well upon what I have said upon the subject, it being a vital question, affecting, as it does, the well-being and the well-doing of her child.

144. *But, if a child's bowels be very costive, what is to be done to relieve them ?*

Do not give him a grain or a drop of opening medicine, but in lieu thereof, administer, by means of a 6 oz. India-rubber Enema Bottle, half a tea-cup or a tea-cupful, according to the age of the child,* of warm water; now this will effectually open the bowels, without confining them afterwards, which opening physic would most assuredly do !

145. *Is it necessary to give a child luncheon ?*

If he want anything to eat between breakfast and dinner let him have a piece of *dry* bread ; and if he have eaten very heartily at dinner, and, like Oliver Twist, " asks for more ! " give him, to satisfy his craving, a piece of *dry* bread. He will never eat more of that than will do him good, and yet he will take sufficient to satisfy his hunger, which is very important.

146. *What ought now to be his dinner ?*

He should now have meat, either mutton or beef, daily, which must be cut up very small, and should be mixed with mealy, *mashed* potato and gravy. He ought *always* to be accustomed to eat salt with his dinner. Let a

---

* For a babe, from birth until he be two years old, one, two, or three table-spoonfuls of warm water will be sufficient, and a 2 oz. Enema Bottle will be the proper size for the purpose of administering it.

mother see that this advice is followed, 'or evil consequences will inevitably ensue. Let him be closely watched,
to ascertain that he well masticates his food, and that he
does not eat too quickly ; for young children are apt to
bolt their food.

147. *Have you any objection to pork for a change ?*

I have a great objection to it for the young. It is a
rich, gross, and therefore unwholesome food for the
delicate stomach of a child. I have known it, in several
instances, produce violent pain; sickness, purging, and
convulsions. If a child be fed much upon such meat, it
will be likely to produce " breakings-out " on the skin.
In fine, his blood will put on the same character as the
food he is fed with. Moreover, pork might be considered
a *strong* meat, and " *strong* meat and *strong* drink can
only be taken by *strong* men."

148. *Do you approve of veal for a child ?*

My objection to pork was, that it was rich and gross ;
this does not apply to veal ; but the objection to it is,
that it is more difficult of digestion that either mutton
or beef; indeed, all young meats are harder of digestion
than meats of maturity ; thus' mutton is more digestible
than lamb, and beef than veal.

149. *Do you disapprove of salted and boiled beef for
a child ?*

If beef be *much* salted it is hard of digestion, and
therefore ought not to be given to him ; but if it have
been but *slightly* salted, then for a change there will be
no objection to a little. There is no necessity in the
winter time to *salt* meat intended for boiling ; then boiled
unsalted meat makes a nice change for a child's dinner.
Salt, of course, *must* with the unsalted meat be eaten.

150. *But suppose there is nothing on the table that a
child may with impunity eat ?*

He should then have either a grilled mutton chop, or
a lightly-boiled egg ; indeed, the latter, at any time,
makes an excellent change. There is great nourishment
in an egg ; it will not only strengthen the frame, but it
will give animal heat as well : these two qualities of an

egg are most valuable ; indeed, essential for the due performance of health : many articles of food contain the
one qualification, but not the other : hence the egg is
admirably suitable for a child's *occasional* dinner.

151. *Are potatoes an unwholesome food for a child ?*

New ones are ; but old potatoes well cooked and mealy,
are the best vegetable he can have. They ought to be
well mashed, as I have known lumps of potatoes cause
convulsions.

152. *Do you approve of any other vegetables for a child?*

Occasionally : either asparagus or broccoli, or cauliflower, or turnips, or French beans, which latter should
be cut up fine, may with advantage be given. Green
peas may occasionally be given, provided they be
thoroughly well boiled, and mashed with the knife on
the plate. Underdone and unmashed peas are not fit .
for a child's stomach : there is nothing more difficult of
digestion than underdone peas. It is important, too, to
mash them, even if they be well done, as a child generally
bolts peas whole ; and they pass through the alimentary
canal without being in the least digested.

153. *Might not a mother be too particular in dieting
her child ?*

Certainly not. If blood can be too pure and too good
she might ! When we take into account that the food
we eat is converted into blood ; that if the food be good
the blood is good ; and that if the food be improper or
impure, the blood is impure likewise ; and, moreover,
when we know that every part of the body is built up
by the blood, we cannot be considered to be too
particular in making our selection of food. Besides if
indigestible or improper food be taken into the stomach,
the blood will not only be made impure, but the stomach
and the bowels will be disordered. Do not let me be
misunderstood : I am no advocate for a child having the
same food one day as another—certainly not. Let there
be variety, but let it be *wholesome* variety. Variety in a
child's (not in infant's) food is necessary. If he were
fed, day after day, on mutton, his stomach would at

length be brought into that state, that in time it would not properly digest any other meat, and a miserable existence would be the result.

154. *What ought a child to drink with his dinner ?*

Toast and water, or, if he prefer it, plain spring water. Let him have as much as he likes. If you give him water to drink, there is no fear of his taking too much ; Nature will tell him when he has had enough. Be careful of the quality of the water, and the source from which you procure it. If the water be *hard*—provided it be free from organic matter—so much the better.* Spring water from a moderately deep well is the best. If it come from a land spring, it is apt, indeed, is almost sure to be contaminated by drains, &c. ; which is a frequent cause of fevers, of diphtheria, of Asiatic cholera, and of other blood poisons.

Guard against the drinking water being contaminated with lead ; never, therefore, allow the water to be collected in leaden cisterns, as it sometimes is if the water be obtained from Water-works companies. Lead pumps, for the same reason, ought never to be used for drinking purposes. Paralysis, constipation, lead colic, dropping of the wrist, wasting of the ball of the thumb, loss of memory, and broken and ruined health, might result from neglect of this advice.

The drinking fountains are a great boon to poor children, as water and plenty of it, is one of the chief necessaries of their existence ; and, unfortunately, at their own homes they are not, oftentimes, able to obtain a sufficient supply. Moreover, drinking fountains are the best advocates for Temperance.

Some parents are in the habit of giving their children beer with their dinners—making them live as they live themselves ! This practice is truly absurd, and fraught with great danger ! not only so, but it is inducing a child to be fond of that which in after life might be his bane

---

* See the *third* edition of *Counsel to a Mother*, under the head of " Hard or soft water as a beverage ? "

and curse! No good end can be obtained by it; it will *not* strengthen so young a child; it will on the contrary, create fever, and will thereby weaken him; it will act injuriously upon his delicate, nervous, and vascular systems, and by means of producing inflammation either of the brain or of its membranes, might thus cause water on the brain (a disease to which young children are subject), or it might induce inflammation of the lungs.

155. *What ought a child who has cut his teeth to have for his supper?*

The same that he has for breakfast. He should sup at six o'clock.

156. *Have you any general remarks to make on a child's meals?*

I recommended a great sameness in an *infant's* diet; but a *child's* meals, his dinners especially, ought to be much varied. For instance, do not let him have day after day mutton; but ring the changes on mutton, beef, poultry, game, and even occasionally fish—sole or cod.

Not only let there be a change of meat, but let there be a change in the manner of cooking it; let the meat sometimes be roasted; let it at other times be boiled. I have known a mother who has prided herself as being experienced in these matters, feed her child, day after day, on mutton chops! Such a proceeding is most injurious to him, as after a while his unfortunate stomach will digest nothing but mutton chops, and, in time, not even those!

With regard to vegetables, potatoes—*mashed* potatoes—ought to be his staple vegetable; but, every now and then, cauliflower, asparagus, turnips, and French beans, should be given.

With respect to puddings, vary them; rice, one day; suet, another; batter, a third; tapioca, a fourth; or, even occasionally, he might have either apple or gooseberry or rhubarb pudding—provided the pastry be plain and light.

It is an excellent plan, as I have before remarked, to let her child eat jam—such as strawberry, raspberry, or gooseberry—and that without stint, either with rice or with batter puddings.

Variety of diet, then, is *good for a child :* it will give him muscle, bone, and sinew ; and, what is very important, it will tend to regulate his bowels, and it will thus prevent the necessity of giving him aperients.

But do not stuff a child—do not press him, as is the wont of some mothers, to eat more than he feels inclined. On the contrary, if you think that he is eating too much—that he is overloading his stomach—and if he should ask for more, then, instead of giving him either more meat or more pudding, give him a piece of dry bread. By doing so, you may rest assured that he will not eat more than is absolutely good for him.

157. *If a child be delicate, is there any objection to a little wine, such as cowslip or tent, to strengthen him ?*

Wine ought not to be given to a child unless it be ordered by a medical man : it is even more injurious than beer. Wine, beer, and spirits, principally owe their strength to the alcohol they contain ; indeed, nearly *all* wines are *fortified* (as it is called) with brandy. Brandy contains a large quantity of alcohol, more than any other liquor, namely 55·3 per cent. If, therefore, you give wine, it is, in point of fact, giving diluted brandy— diluted alcohol ; and alcohol acts, unless it be used as a medicine, and under skilful medical advice, as a poison to a child.

158. *Suppose a child suddenly to lose his appetite ? is any notice to be taken of it ?*

If he cannot eat well, depend upon it, there is something wrong about the system. If he be teething, let a mother look well to his gums, and satisfy herself that they do not require lancing. If they be red, hot, and swollen, send for a medical man, that he may scarify them. If his gums be not inflamed, and no tooth appears near, let her look well to the state of his bowels ; let her ascertain that they be sufficiently opened, and that the stools be of a proper consistence, colour, and smell. If they be neither the one nor the other, give a dose of aperient medicine, which will generally put all to rights. If the gums be cool, and the bowels be right, and his appetite continue bad, call in medical aid.

A child asking for something to eat, is frequently, in a severe illness, the first favourable symptom ; we may generally then prognosticate that all will soon be well again.

If a child refuse his food, neither coax nor tempt him to eat : as food without an appetite will do him more harm than it will do him good ; it may produce either sickness, bowel-complaint, or fever. Depend upon it, there is always a cause for a want of appetite ;—perhaps his stomach has been over-worked, and requires repose ; or his bowels are loaded, and Nature wishes to take time to use up the old material ;—there might be fever lurking in his system ; Nature stops the supplies, and thus endeavours, by not giving it food to work with, to nip it in the bud ;—there might be inflammation ; food would then be improper, as it would only add fuel to the fire ; let, therefore, the cause be either an over-worked stomach, over-loaded bowels, fever, or inflammation, food would be injurious. Kind Nature if we will but listen to her voice, will tell us when to eat, and when to refrain.

159. *When a child is four or five years old, have you any objection to his drinking tea ?*

Some parents are in the habit of giving their children strong (and frequently green) tea. This practice is most hurtful. It acts injuriously upon their delicate, nervous system, and thus weakens their whole frame. If milk does not agree, a cup of very weak tea, that is to say, water with a dash of *black* tea in it, with a table-spoonful of cream, may be substituted for milk ; but a mother must never give tea where milk agrees.

160. *Have you any objection to a child occasionally having either cakes or sweetmeats ?*

I consider them as so much slow poison. Such things both cloy and weaken the stomach, and thereby take away the appetite, and thus debilitate the frame. Moreover "sweetmeats are coloured with poisonous pigments." A mother, surely, is not aware, that when she is giving her child Sugar Confectionery she is, in many cases, administering a deadly poison to him? "We beg to direct the attention of our readers to the Report of the Analytical Sanitary Commission, contained in the *Lancet* of the present week (Dec. 18, 1858), on the pigments employed in colouring articles of Sugar Confectionery. From this report it appears that metallic pigments of a highly dangerous and even poisonous character, containing chromic acid, lead, copper, mercury, and arsenic, are commonly used in the colouring of such articles."

If a child be never allowed to eat cakes and sweetmeats, he will consider a piece of dry bread a luxury, and will eat it with the greatest relish.

161. *Is bakers' or is home-made bread the most wholesome for a child ?*

Bakers' bread is certainly the lightest ; and, if we could depend upon its being unadulterated, would, from its lightness, be the most wholesome ; but as we cannot always depend upon bakers' bread, home-made bread, as a rule should be preferred. If it be at all heavy, a child must not be allowed to partake of it ; a baker's loaf ought then to be sent for, and continued to be eaten until light home-made bread can be procured. Heavy bread is most indigestible. He must not be allowed to eat bread until it be two or three days old. If it be a week old, in cold weather, it will be the more wholesome.

162. *Do you approve either of caraway seeds or of currants in bread or in cakes—the former to disperse wind, the latter to open the bowels ?*

There is nothing better than plain bread : the caraway-seeds generally pass through the bowels undigested, and thus might irritate, and might produce, instead of disperse wind.* Some mothers put currants in cakes, with

---

* Although caraway seeds *whole* are unwholesome, yet caraway-

a view of opening the bowels of their children ; but they only open them by disordering them.

163. *My child has an antipathy to certain articles of diet : what would you advise to be done ?*

A child's antipathy to certain articles of diet should be respected : it is a sin and a shame to force him to eat what he has a great dislike to : a child, for instance, sometimes dislikes the fat of meat, underdone meat, the skin off boiled milk and off rice-pudding. Why should he not have his likes and dislikes as well as "children of a larger growth ?" Besides, there is an idiosyncrasy —a peculiarity of the constitution in some children—and Nature oftentimes especially points out what is good and what is bad for them individually, and we are not to fly in the face of Nature. "What is one man's meat is another man's poison." If a child be forced to eat what he dislikes, it will most likely not only make him sick, but will disorder his stomach and bowels : food, if it is really to do him good, must be eaten by him with a relish, and not with disgust and aversion. Some mothers, who are strict disciplinarians, pride themselves on compelling their children to eat whatever they choose to give them ! Such children are to be pitied !

164. *When ought a child to commence to dine with his parents ?*

As soon as he be old enough to sit up at the table, provided the father and mother either dine or lunch in the middle of the day. "I always prefer having children about me at meal times. I think it makes them little gentlemen and gentlewomen in a manner that nothing else will."—*Christian's Mistake.*

THE NURSERY.

165. *Have you any remarks to make on the selection, the ventilation, the warming, the temperature, and the arrangements of a nursery ? and have you any further*

---

tea, made as recommended in a previous Conversation, is an excellent remedy to disperse wind.

H

observations to offer conducivce to the well-doing of my child ?

The nursery ought to be the largest and the most airy room in the house. In the town, if it be in the topmost story (provided the apartment be large and airy) so much the better, as the air will then be purer. The architect, in the building of a house, ought to be particularly directed to pay attention to the space, the loftiness, the ventilation, the light, the warming, and the conveniences of a nursery. A bath-room attached to it will be of great importance and benefit to the health of a child.

It will be advantageous to have a water-closet near at hand, which should be well supplied with water, be well drained, and be well ventilated. If this be not practicable, the evacuations ought to be removed as soon as they are passed. It is a filthy and an idle habit of a nurse-maid to allow a motion to remain for any length of time in the room.

The VENTILATION of a nursery is of paramount importance. There ought to be a constant supply of fresh pure air in the apartment. But how few nurseries have fresh, pure air ! Many nurseries are nearly hermetically sealed —the windows are seldom, if ever, opened ; the doors are religiously closed ; and, in summer time, the chimneys are carefully stuffed up, so that a breath of air is not allowed to enter ! The consequences are, the poor unfortunate children " are poisoned by their own breaths," and are made so delicate that they are constantly catching cold ; indeed, it might be said that they are labouring under chronic catarrhs, all arising from Nature's laws being set at defiance.

The windows ought to be large, and should be made to freely open both top and bottom. Whenever the child is out of the nursery, the windows ought to be thrown wide open ; indeed, when he is in it, if the weather be fine, the upper sash should be a little lowered. A child should be encouraged to change the room frequently, in order that it may be freely ventilated ; for good air is as necessary to his health as wholesome food, and air cannot be good if it be not frequently changed. If you wish to have a strong and healthy child, ponder over and follow this advice.

I have to enter my protest against the use of a stove in a nursery. I consider a gas stove *without a chimney* to be an abomination, most destructive to human life. There is nothing like the old-fashioned open fire-place with a good-sized chimney, so that it may not only carry off the smoke, but also the impure air of the room.

Be strict in not allowing your child either to touch or to play with fire; frightful accidents have occurred from mothers and nurses being on these points lax. The nursery ought to have a large fire-guard, to go all round the hearth, and which should be sufficiently high to prevent a child from climbing over. Not only must the nursery have a guard, but every room where he is allowed to go should be furnished with one on the bars.

Moreover, it will be advisable to have a guard in every room where a fire is burning, to prevent ladies from being burned. Fortunately for them, preposterous crinolines are out of fashion : when they were in fashion, death from burning was of every-day occurrence ; indeed, lady-burning was then to be considered one of the institutions of our land !

A nursery is usually kept too hot; the temperature in the winter time ought *not to exceed* 60 degrees Fahrenheit. A *good* thermometer should be considered an indispensable requisite to a nursery. A child in a hot, close nursery is bathed in perspiration ; if he leave the room to go to one of lower temperature, the pores of his skin are suddenly closed, and either a severe cold or an inflammation of the lungs, or an attack of bronchitis, is likely to ensue. Moreover, the child is both weakened and enervated by the heat, and thus readily falls a prey to disease.

A child ought never to be permitted to sit with his back to the fire ; if he be allowed, it weakens the spine, and thus his whole frame ; it causes a rush of blood to the head and face, and predisposes him to catch cold.

Let a nurse make a point of opening the nursery window every time that she and her little charge leave the nursery, if her absence be only for half an hour. The mother herself ought to see that this advice is followed, pure air is so essential to the well-being of a child. Pure air and pure water, and let me add, pure milk, are for a child the grand and principal requirements of health.

Look well to the DRAINAGE of your house and neighbourhood. A child is very susceptible to the influence of bad drainage. Bad drains are fruitful sources of scarlet fever, of diphtheria, of diarrhœa, &c. " It is sad to be reminded that, whatever evils threaten the health of population, whether from pollutions of water or of air,—whether from bad drainage or overcrowding, they fall heaviest upon the most innocent victims—upon children of tender years. Their delicate frames are infinitely more sensitive than the hardened constitutions of adults, and the breath of poison, or the chill of hardships, easily blights their tender life."—*The Times.*

A nursery floor ought not to be *washed* oftener than once a week; and then the child or children should, until it be dry, be sent into another room. During the drying of the floor, the windows must, of course, be thrown *wide* open.

The constant *wetting* of a nursery is a frequent source of illness among children. The floor ought, of course, to be kept clean; but this may be done by the servant thoroughly sweeping the room out every morning before her little charge makes his appearance.

Do not have your nursery wall covered with *green* paper-hangings. Green paper-hangings contain large quantities of arsenic—arsenite of copper (Scheele's green)—which, I need scarcely say, is a virulent poison, and which flies about the room in the form of powder. There is frequently enough poison on the walls of a room to destroy a whole neighbourhood.

There is another great objection to having your nursery walls covered with *green* paper-hangings; if any of the paper should become loose from the walls, a little child is very apt to play with it, and to put it, as he does every thing else, to his mouth. This is not an imaginary state of things, as four children in one family have just lost their lives from sucking green paper-hangings.

Green dresses, as they are coloured with a preparation of arsenic, are equally as dangerous as green paper-hangings; a child ought, therefore, never to wear a *green* dress. "It may be interesting to some of our readers," says *Land and Water*, "to know that the new green, so fashionable for ladies' dresses, is just as dangerous in its nature as the green wall-paper, about which so much was written some time since. It is prepared with a large quantity of arsenic; and we have been assured by several of the leading dressmakers, that the workwomen employed in making up dresses of this colour are seriously affected with all the symptoms of arsenical poisoning. Let our lady friends take care."

Children's toys are frequently painted of a green colour with arsenite of copper, and are consequently, highly dangerous for him to play with. The best toy for a child is a box of *unpainted* wooden bricks, which is a constant source of amusement to him.

If you have your nursery walls hung with paintings and engravings, let them be of good quality. The horrid daubs and bad engravings that usually disfigure nursery walls, are enough to ruin the taste of a child, and to make him take a disgust to drawing, which would be a misfortune. A fine engraving and a good painting expand and elevate his mind. We all know that first impressions are the most vivid and the most lasting. A taste in early life for everything refined and beautiful purifies his mind, cultivates his intellect, keeps him from low company, and makes him grow up a gentleman !

Lucifer matches, in case of sudden illness, should, both in the nursery and in the bedroom, be always in readiness; but they must be carefully placed out of the reach of children, as lucifer matches are a deadly poison.

Many inquests have been held on children who have, from having sucked them, been poisoned by them.

166. *Have you any observation to make on the* LIGHT *of a nursery?*

Let the window, or what is better, the windows, of a nursery be very large, so as to thoroughly light up every nook and corner of the room, as there is nothing more conducive to the health of a child than an abundance of light in the dwelling. A room cannot, then, be too light. The .windows of a nursery are generally too small. A child requires as much light as a plant. Gardeners are well aware of the great importance of light in the construction of their greenhouses, and yet a child, who requires it as much, and is of much greater importance, is cooped up in dark rooms !

The windows of a nursery ought not only to be frequently opened to let in fresh air, *but should be frequently cleaned,* to let in plenty of light and of sunshine, as nothing is so cheering and beneficial to a child as an abundance of light and sunshine !

With regard to the best artificial light for a nursery. —The air of a nursery cannot be too pure ; I therefore do not advise you to have gas in it, as gas in burning gives off quantities of carbonic acid and sulphuretted hydrogen, which vitiate the air. The paraffine lamp, too, makes a room very hot and close. There is no better light for a nursery than either Price's patent candles or the old-fashioned tallow-candle.

Let a child's *home* be the happiest *house* to him in the world ; and to be happy he must be merry, and all around him should be merry and cheerful ; and he ought to have an abundance of playthings, to help on the merriment. If he have a dismal nurse, and a dismal home, he may as well be incarcerated in a prison, and be attended by a gaoler. It is sad enough to see dismal, doleful men and women, but it is a truly lamentable and unnatural sight to see a doleful child ! The young ought to be as playful and as full of innocent mischief as a kitten. There will be quite time enough in after years for sorrow and for sadness.

Bright colours, plenty of light, *clean* windows (mind this, if you please), an abundance of *good*-coloured prints, and toys without number, are the proper furnishings of a nursery. Nursery ! why, the very name tells you what it ought to be—the home of childhood—the most important room in the house,—a room that will greatly tend to stamp the character of your child for the remainder of his life.

167. *Have you any more hints to offer conducive to the well-doing of my child ?*

You cannot be too particular in the choice of those who are in constant attendance upon him. You yourself, of course, must be his *head-nurse*—you only require some one to take the drudgery off your hands ! You ought to be particularly careful in the selection of his nurse. She should be steady, lively, truthful, and good tempered ; and must be free from any natural imperfection, such as squinting, stammering, &c., for a child is such an imitative creature that he is likely to acquire that defect, which in the nurse is natural. " Children, like babies, are quick at ' taking notice.' What they see they mark, and what they mark they are very prone to copy." —*The Times.*

She ought not to be very young, or she may be thoughtless, careless, and giggling. You have no right to set a child to mind a child ; it would be like the blind leading the blind. No ! a child is too precious a treasure to be entrusted to the care and keeping of a young girl. Many a child has been ruined for life by a careless young nurse dropping him and injuring his spine.

A nurse ought to be both strong and active, in order that her little charge may have plenty of good nursing ; for it requires great strength in the arms to carry a heavy child for the space of an hour or two at a stretch, in the open air ; and such is absolutely necessary, and is the only way to make him strong, and to cause him to cut his teeth easily, and at the same time to regulate his bowels; a nurse, therefore, must be strong and active, and not mind hard work, for hard work it is; but, after she is accustomed to it, pleasant notwithstanding.

Never should a nurse be allowed to wear a mask, nor to dress up and paint herself as a ghost, or as any other frightful object. A child is naturally timid and full of fears, and what would not make the slightest impression upon a grown-up person might throw a child into fits—

> " The sleeping, and the dead,
> Are but as pictures : 'tis the age of childhood
> That fears a painted devil."—*Shakspeare.*

Never should she be permitted to tell her little charge frightful stories of ghosts and hobgoblins; if this be allowed, the child's disposition will become timid and wavering, and may continue so for the remainder of his life.

If a little fellow were not terrified by such stories, the darkness would not frighten him more than the light. Moreover, the mind thus filled with fear, acts upon the body, and injures the health. A child must never be placed in a dark cellar, nor frightened by tales of rats, &c. Instances are related of fear thus induced impairing the intellect for life; and there are numerous examples of sudden fright causing a dangerous and even a fatal illness.

Night-terrors.—This frightening of a child by a silly nurse frequently brings on night-terrors. He wakes up suddenly, soon after going to sleep, frightened and terrified; screaming violently, and declaring that he has seen either some ghost, or thief, or some object that the silly nurse had been previously in the day describing, who is come for him to take him away. The little fellow is the very picture of terror and alarm; he hides his face in his mother's bosom, the perspiration streams down him, and it is some time before he can be pacified— when, at length, he falls into a troubled feverish slumber, to awake in the morning unrefreshed. Night after night these terrors harass him, until his health materially suffers, and his young life becomes miserable looking forward with dread to the approach of darkness.

Treatment of night-terrors.—If they have been brought on by the folly of the nurse, discharge her at once, and be careful to select a more discreet one. When the child retires to rest, leave a candle burning, and let it burn all night ; sit with him until he be asleep ; and take care, in case he should rouse up in one of his night-terrors, that either yourself or some kind person be near at hand. Do not scold him for being frightened—he cannot help it ; but soothe him, calm him, fondle him, take him into your arms and let him feel that he has some one to rest upon, to defend and to protect him. It is frequently in these cases necessary before he can be cured to let him have change of air and change of scene. Let him live, in the day time, a great part of the day in the open air.

A nurse-maid should never, on any account whatever, be allowed to whip a child. "Does ever any man or woman remember the feeling of being 'whipped' as a child, the fierce anger, the insupportable ignominy, the longing for revenge, which blotted out all thought of contrition for the fault or rebellion against the punishment ? With this recollection on their own parts, I can hardly suppose any parents venturing to inflict it, much less allowing its infliction by another under any circumstances whatever. A nurse-maid or domestic of any sort, once discovered to have lifted up her hand against a child, ought to meet instant severe rebuke, and on a repetition of the offence instant dismissal."*

I have seen in the winter time a lazy nurse sit before the fire with a child on her lap, rubbing his cold feet just before putting him to his bed. Now, this is not the way to warm his feet. The right method is to let him romp and run either about the room, or the landing, or the hall—this will effectually warm them ; but, of course, it will entail a little extra trouble on the nurse,

---

* *A Woman's Thoughts about Women.*

as she will have to use a little exertion to induce him to
do so, and this extra trouble a lazy nurse will not relish.
Warming the feet before the fire will give the little
fellow chilblains, and will make him when he is in bed
more chilly. The only way for him to have a good
romp before he goes to bed, is for the mother to join in
the game. She may rest assured, that if she does so,
her child will not be the only one to benefit by it. She
herself will find it of marvellous benefit to her own
health ; it will warm her own feet, it will be almost sure
to insure her a good night, and will make her feel so
light and buoyant as almost to fancy that she is a girl
again ! Well, then, let every child, before going to bed,
hold a high court of revelry, let him have an hour—the
Children's Hour—devoted to romp, to dance, to shout,
to sing, to riot, and to play, and let him be the master
of the revels— .

> " Between the dark and the daylight,
> When the night is beginning to lower,
> Comes a pause in the day's occupation,
> Which is known as the Children's Hour."
> *Longfellow.*

. Let a child be employed—take an interest in his
employment, let him fancy that he is useful—*and he is
useful*, he is laying in a stock of health. He is much
more usefully employed than many other grown-up
children are !

A child should be happy ; he must, in every way, be
made happy ; everything ought to be done to conduce to
his happiness, to give him joy, gladness, and pleasure.
Happy he should be, as happy as the day is long.
Kindness should be lavished upon him. Make a child
understand that you love him ; prove it in your
actions—these are better than words ; look after his little
pleasures—join in his little sports ; let him never hear a
morose word—it would rankle in his breast, take deep
root, and in due time bring forth bitter fruit. Love !
let love be his pole-star ; let it be the guide and the rule
of all you do and all you say unto him. Let your face,

as well as your tongue speak love. Let your hands be ever ready to minister to his pleasures and to his play. " Blessed be the hand that prepares a pleasure for a child, for there is no saying when and where it may again bloom forth. Does not almost everybody remember some kind-hearted man who showed him a kindness in the dulcet days of childhood? The writer of this recollects himself, at this moment, a bare-footed lad, standing at the wooden fence of a poor little garden in his native village, while, with longing eyes, he gazed on the flowers which were blooming there quietly in the brightness of the Sabbath morning. The possessor came from his little cottage. He was a wood-cutter by trade, and spent the whole week at work in the woods. He had come into the garden to gather flowers to stick in his coat when he went to church. He saw the boy, and breaking off the most beautiful of his carnations (it was streaked with red and white), he gave it to him. Neither the giver nor the receiver spoke a word, and with bounding steps the boy ran home. And now, here, at a vast distance from that home, after so many events of so many years, the feeling of gratitude which agitated the breast of the boy, expressed itself on paper. The carnation has long since faded, but it now bloometh afresh."—*Douglas Jerrold.*

The hearty ringing laugh of a child is sweet music to the ear. There are three most joyous sounds in nature— the hum of a bee, the purr of a cat, and the laugh of a child. They tell of peace, of happiness, and of content- ment, and make one for a while forget that there is so much misery in the world.

A man who dislikes children is unnatural : he has no " milk of human kindness " in him ; he should be shunned. Give me, for a friend, a man—

" Who takes the children on his knee,
And winds their curls, about his hand."—*Tennyson.*

168. *If a child be peevish, and apparently in good health, have you any plan to propose to allay his irrita- bility?*

A child's troubles are soon over—his tears are soon dried; "nothing dries sooner than a tear"—if not prolonged by improper management—

> "The tear down childhood's cheek that flows
> Is like the dew-drop on the rose;
> When next the summer breeze comes by,
> And waves the bush, the flower is dry."—*Scott.*

Never allow a child to be teased; it spoils his temper. If he be in a cross humour take no notice of it, but divert his attention to some pleasing object. This may be done without spoiling him. Do not combat bad temper with bad temper—noise with noise. Be firm, be kind, be gentle,* be loving, speak quietly, smile tenderly, and embrace him fondly, but *insist upon implicit obedience*, and you will have, with God's blessing, a happy child—

> "When a little child is weak
> From fever passing by,
> Or wearied out with restlessness
> Don't scold him if he cry.
>
> Tell him some pretty story—
> Don't read it from a book;
> He likes to watch you while you speak,
> And take in every look.
>
> Or sometimes singing gently—
> A little song may please,
> With quiet and amusing words,
> And tune that flows with ease.
>
> Or if he is impatient,
> Perhaps from time to time
> A simple hymn may suit the best,
> In short and easy rhyme.
>
> The measured verses flowing
> In accents clear and mild,
> May blend into his troubled thought,
> And soothe the little child.
>
> But let the words be simple,
> And suited to his mind,
> And loving, that his weary heart
> A resting-place may find."—*Household Verses.*

* "But we were gentle among you, even as a women cherisheth her children."—1 Thess. ii. 7.

Speak *gently* to a child ; speak *gently* to all ; but more especially speak *gently* to a child. "A gentle voice is an excellent thing in a woman," and is a jewel of great price, and is one of the concomitants of a *perfect* lady. Let the hinges of your disposition be well oiled. " ' I have a dear friend. He was one of those well-oiled dispositions which turn upon the hinges of the world without creaking.' Would to heaven there were more of them ! How many there are who never turn upon the hinges of this world without a grinding that sets the teeth of a whole household on edge ! And somehow or other it has been the evil fate of many of the best spirits to be so circumstanced ; both men and women, to whom life is ' sweet habitude of being,' which has gone far to reconcile them to solitude as far less intolerable ! To these especially the creakings of those said rough hinges of the world is one continued torture, for they are all too finely strung ; and the oft-recurring grind jars the whole sentient frame, mars the beautiful lyre, and makes cruel discord in a soul of music. How much of sadness there is in such thoughts ! Seems there not a Past in some lives, to which it is impossible ever to become reconciled !"—*Life's Problems.*

Pleasant words ought always to be spoken to a child ; there must be neither snarling, nor snapping, nor snubbing, nor loud contention towards him. If there be it will ruin his temper and disposition, and will make him hard and harsh, morose and disagreeable.

Do not always be telling your child how wicked he is ; what a naughty boy he is ; that God will never love him, and all the rest of such twaddle and blatant inanity ! Do not, in point of fact, bully him, as many poor little fellows are bullied ! It will ruin him if you do ; it will make him in after years either a coward or a tyrant. Such conversations, like constant droppings of water, will make an impression, and will cause him to feel that it is of no use to try to be good—that he is hopelessly wicked ! Instead of such language, give him confidence in himself ; rather find out his good points

and dwell upon them ; praise him where and whenever
you can ; and make him feel that, by perseverance and
by God's blessing, he will make a good man. Speak
truthfully to your child ; if you once deceive him, he
will not believe you for the future. Not only so, but if
you are truthful yourself you are likely to make him
truthful—like begets like.   There is something beautiful in truth ! A lying child is an abomination ! Sir
Walter Scott says "that he taught his son to ride, to
shoot, and to tell the truth."   Archdeacon Hare asserts
" that Purity is the feminine, Truth the masculine of
Honour."

As soon as a child can speak he should be made to
lisp the noble words of truth, and to .love it, and to
abhor a lie ! What a beautiful character he will then
make ! Blessed is the child that can say,—

> " Parental cares watched o'er my growing youth,
> And early stamped it with the love of truth."
> *Leadbeater Papers.*

Have no favourites, show no partiality ; for the young
are very jealous, sharp-sighted, and quick-witted, and
take a dislike to the petted one.   Do not rouse the old
Adam in them.   Let children be taught to be " kindly
affectioned one to another with brotherly love ;" let
them be encouraged to share each other's toys and playthings, and to banish selfishness.

Attend to a child's *little* pleasures.   It is the *little*
pleasures of a child that constitute his happiness.   Great
pleasures to him and to us all (as a favourite author
remarks) come but seldom, and are the exceptions, and
not the rule.

Let a child be nurtured in love.   " It will be seen,"
says the author of *John Halifax*, " that I hold this law
of kindness as the Alpha and Omega of education.   I
once asked one, in his own house, a father in everything
but the name. his authority unquestioned, his least word
held in reverence, his smallest wish obeyed—' How did
you ever manage to bring up these children ?' He said
By love.' "

Let every word and action prove that you love your children. Enter into all their little pursuits and pleasures. Join them in their play, and be a "child again !" If they are curious, do not check their curiosity ; but rather encourage it ; for they have a great deal—as we all have—to learn, and how can they know if they are not taught? You may depend upon it the knowledge they obtain from observation is far superior to that obtained from books. Let all you teach them, let all you do, and let all you say bear the stamp of love. "Endeavour, from first to last, in your intercourse with your children, to let it bear the impress of *love*. It is not enough that you *feel* affection towards your children—that you are devoted to their interests ; you must show in your manner the fondness of your hearts towards them. Young minds cannot appreciate great sacrifices made for them ; they judge their parents by the words and deeds of every-day life. They are won by *little* kindnesses, and alienated by *little* acts of neglect or impatience. One complaint unnoticed, one appeal unheeded, one lawful request arbitrarily refused, will be remembered by your little ones more than a thousand acts of the most devoted affection."—*The Protoplast.*

A placid, well-regulated temper is very conducive to health. A disordered, or an over-loaded stomach, is a frequent cause of peevishness. Appropriate treatment in such a case will, of course, be necessary.

169. *My child stammers : can you tell me the cause, and can you suggest a remedy ?*

A child who stammers is generally " nervous," quick, and impulsive. His ideas flow too rapidly for speech. He is "nervous ; " hence, when he is alone, and with those he loves, he oftentimes speaks fluently and well ; he stammers more both when he is tired and when he is out of health—when the nerves are either weak or exhausted. He is emotional : when he is either in a passion or in excitement, either of joy or of grief, he can scarcely speak—"he stammers all over." He is impulsive ; he often stammers in consequence. He is in too

great a hurry to bring out his words ; they do not flow
in proper sequence : hence his words are broken and
disjointed.

Stammering, of course, might be owing either to some
organic defect, such as from defective palate, or from
defective brain, then nothing will cure him ; or it might
be owing to " nervous " causes—to " irregular nervous
action," then a cure might, with care and perseverance,
be usually effected.

In all cases of stammering of a child, let both the
palate of his mouth and the bridle of his tongue be carefully examined, to see that neither the palate be defective,
nor the bridle of the tongue be too short—that he be
not tongue-tied.

Now, with regard to Treatment.—Make him speak
slowly and deliberately ; let him form each word, without clipping or chopping ; let him be made, when you
are alone with him, to exercise himself in elocution. If
he speak quickly, stop him in his mid-career, and make
him, quietly and deliberately, go through the sentence
again and again, until he has mastered the difficulty ;
teach him to collect his thoughts, and to weigh each word
ere he give it utterance ; practise him in singing little
hymns and songs for children ; this you will find a valuable help in the cure. A stammerer seldom stutters when
he sings. When he sings, he has a full knowledge of the
words, and is obliged to keep in time—to sing neither
too fast nor too slow. Besides, he sings in a different
key to his speaking voice. Many professors for the
treatment of stammering cure their patients by practising
lessons of a sing-song character.

Never jeer him for stammering, nor turn him to
ridicule ; if you do, it will make him ten times worse ;
but be patient and gentle with him, and endeavour to
give him confidence, and encourage him to speak to you
as quietly, as gently, and deliberately as you speak to
him ; tell him not to speak until he has arranged his
thoughts and chosen his words ; let him do nothing in a
hurry

Demosthenes was said, in his youth, to have stammered fearfully, and to have cured himself by his own prescription, namely, by putting a pebble in his mouth, and declaiming, frequently, slowly, quietly, and deliberately, on the sea-shore—the fishes alone being his audience,—until at length he cured himself, and charmed the world with his eloquence and with his elocution. He is held up, to this very day, as the personification and as the model of an orator. His patience, perseverance, and practice ought, by all who either are, or are interested in a stammerer, to be borne in mind and followed.

170. *Do you approve of a carpet in a nursery ?*

No ; unless it be a small piece for a child to roll upon. A carpet harbours dirt and dust, which dust is constantly floating about the atmosphere, and thus making it impure for him to breathe. The truth of this may be easily ascertained by entering a darkened room, where a ray of sunshine is struggling through a crevice in the shutters. If the floor of a nursery must be covered, let drugget be laid down ; and this may every morning be taken up and shaken. The less furniture a nursery contains the better ; for much furniture obstructs the free circulation of the air, and, moreover, prevents a child from taking proper play and exercise in the room—an abundance of which are absolutely necessary for his health.

171. *Supposing there is not a fire in the nursery grate, ought the chimney to be stopped to prevent a draught in the room ?*

Certainly not. I consider the use of a chimney to be two-fold :—first, to carry off the smoke ; and secondly (which is of quite as much importance), to ventilate the room, by carrying off the impure air, loaded as it is with carbonic acid gas—the refuse of · respiration. The chimney, therefore, should never, either winter or summer, be allowed for one moment to be stopped. This is important advice, and requires the strict supervision of every mother, as servants will, if they have the chance, stop all chimneys that have no fires in the grates.

I

EXERCISE.

172. *Do you approve, during the summer months, of sending a child out* BEFORE *breakfast ?*

I do, when the weather will permit, and provided the wind be neither in an easterly nor in a north-easterly direction ; indeed, *he can scarcely be too much in the open air.* He must not be allowed to stand about draughts or about entries, and the only way to prevent him doing so is for the mother herself to accompany the nurse. She will then kill two birds with one stone, as she will, by doing so, benefit her own as well as her child's health.

173. *Ought a child to be early put on his feet to walk ?*

No : let him learn to walk himself. He ought to be put upon a carpet; and it will be found that when he is strong enough, he will hold by a chair, and will stand alone : when he can do so, and attempts to walk, he should then be supported. You must, on first putting him upon his feet, be guided by his own wishes. He will, as soon as he is strong enough to walk, have the inclination to do so. When he has the inclination and the strength it will be folly to restrain him.; if he have neither the inclination nor the strength, it will be absurd to urge him on. Rely, therefore, to a certain extent, upon the inclination of the child himself. Self-reliance cannot be too early taught him, and, indeed, every one else. In the generality of instances, however, a child is put on his feet too soon, and the bones, at that tender age, being very flexible, bend, causing bowed and bandy-legs ; and the knees, being weak, approximate too closely together, and thus they become knock-kneed. This advice of *not* putting a child *early* on his feet, I must strongly insist on, as many mothers are so ridiculously ambitious that their young ones should walk early—that they should walk before other children of their acquaintance have attempted—that they have frequently caused the above lamentable deformities ; which is a standing reproach to them during the rest of their lives !

174. *Do you approve of perambulators ?*

I do not, for two reasons :—first, because when a child is strong enough, he had better walk as much as he will ; and, secondly, the motion is not so good, and the muscles are not so much put into action, and consequently cannot be so well developed, as when he is carried. A perambulator is very apt to make a child stoop, and to make him both crooked and round-shouldered. He is cramped by being so long in one position. It is painful to notice a babe of a few months old in one of these new-fangled carriages. His little head is bobbing about first on one side and then on the other—at one moment it is dropping on his chest, the next it is forcibly jolted behind : he looks, and doubtless feels, wretched and uncomfortable. Again, these perambulators are dangerous in crowded thoroughfares. They are a public nuisance, inasmuch as they are wheeled against and between people's legs, and are a fruitful source of the breaking of shins, of the spraining of ankles, of the crushing of corns, and of the ruffling of the tempers of the foot-passengers who unfortunately come within their reach ; while, in all probability, the gaping nurses are staring another way, and every way indeed but the right, more especially if there be a redcoat in the path !

Besides, in very cold weather, or in a very young infant, the warmth of the nurse's body, while he is being carried, helps to keep him warm, he himself being naturally cold. In point of fact, the child, while being borne in the nurse's arms, reposes on the nurse, warm and supported, as though he were in a nest ! While, on the other hand, if he be in a perambulator, he is cold and unsupported, looking the very picture of misery, seeking everywhere for rest and comfort, and finding none !

A nurse's arm, then, is the only proper carriage for a *young* child to take exercise on. She ought to change about, first carrying him on the one arm, and then on the other. Nursing him on one arm only might give his body a twist on one side, and thus might cause deformity.

When he is old enough to walk, and is able properly ,

to support the weight of his own neck and back, then there will be no objection, provided it be not in a crowded thoroughfare, to his riding occasionally in a perambulator; but when he is older still, and can sit either a donkey or a pony, such exercise will be far more beneficial, and will afford him much greater pleasure.

175. *Supposing it to be wet under foot, but dry above, do you then approve of sending a child out ?*

If the wind be neither in the east nor the north-east, and if the air be not damp, let him be well wrapped up and be sent out. If he be labouring under an inflammation of the lungs, however slight, or if· he be just recovering from one, it would, of course, be highly improper. In the management of a child, we must take care neither to coddle nor to expose him unnecessarily, as both are dangerous.

Never send a child out to walk in a fog; he will, if you do, be almost sure to catch cold. It would be much safer to send him out in rain than in fog, though neither the one nor the other would be desirable.

176. *How many times a day in fine weather ought a child to be sent out ?*

Let him be sent out as often as it be possible. If a child lived more in the open air than he is wont to do, he would neither be so susceptible of disease, nor would he suffer so much from teething, nor from catching cold.

177. *Supposing the day to be wet, what exercise would you then recommend ?*

The child ought to run either about a large room, or about the hall ; and if it does not rain violently, you should put on his hat and throw up the window, taking care while the window is open that he does not stand still. A wet day is the day for him to hold his high court of revelry, and "to make him as happy as the day is long."

Do not on any account allow him to sit any length of time at a table, amusing himself with books, &c. ; let him be active and stirring, that his blood may freely circulate as it ought to do, and that his muscles may be well developed. I would rather see him actively engaged in mischief than sitting still, doing nothing ! He ought to be put on the carpet, and should then be tumbled and rolled about, to make the blood bound merrily through the vessels, to stir up the liver, to promote digestion, and to open the bowels. The misfortune of it is, the present race of nurses are so encumbered with long dresses, and so screwed in with tight stays (aping their betters), that they are not able to stoop properly, and thus to have a good game of romps with their little charges. " Doing nothing is doing ill " is as true a saying as was ever spoken.

178. *Supposing it to be winter, and the weather to be very cold, would you still send a child out ?*

Decidedly, provided he be well wrapped up. The cold will brace and strengthen him. Cold weather is the finest tonic in the world.

In frosty weather, the roads being slippery, when you send him out to walk, put a pair of large old woollen stockings *over* his boots or shoes. This will not only keep his feet and his legs warm, but it will prevent him from falling down and hurting himself. While thus equipped, he may even walk on a slide of ice without falling down !

A child, in the winter time, requires, to keep him warm, plenty of flannel and plenty of food, plenty of fresh and genuine milk, and plenty of water in his tub to wash and bathe him in a morning, plenty of exercise and plenty of play, and then he may brave the frosty air. It is the coddled, the half-washed, and the half-starved child (half-washed and half-starved from either the mother's ignorance or from the mother's timidity), that is the chilly starveling,—catching cold at every breath of wind, and every time he either walks or is carried out,—a puny, skinny, scraggy, scare-crow, more dead than alive, and more fit for his grave than for the rough world he will have to struggle in ! If the above advice be strictly followed, a child may be sent out in the coldest weather, even—

> " When icicles hang by the wall,
>   And Dick, the shepherd, blows his nail ;
> And Tom bears logs into the hall,
>   And milk comes frozen home in pail."
>
>   *Shakspeare.*

AMUSEMENTS.

179. *Have you any remarks to make on the amusements of a child ?*

Let the amusements of a child be as much as possible out of doors; let him spend the greater part of every day in the open air ; let him exert himself as much as he please, his feelings will tell him when to rest and when to begin again ; let him be what Nature intended him to be—a happy, laughing, joyous child. Do not let him be always poring over books :—

> " Books ! 'tis a dull and endless strife,
>   Come, hear the woodland linnet !
> How sweet his music ! On my life,
>   There's more of wisdom in it.
>
> And hark ! how blithe the throstle sings !
>   He, too, is no mean preacher :
> Come forth into the light of things,—
>   Let Nature be your teacher.
>
> She has a world of ready wealth,
>   Our minds and hearts to bless,—
> Spontaneous wisdom breathed by health,
>   Truth breathed by cheerfulness.
>
> One impulse from a vernal wood
>   May teach you more of man,
> Of moral evil and of good,
>   Than all the sages can."—*Wordsworth.*

He ought to be encouraged to engage in those sports wherein the greatest number of muscles are brought into play. For instance, to play at ball, or hoop, or football, to play at horses, to run to certain distances and back ; and, if a girl, to amuse herself with a skipping rope, such being excellent exercise—

> " By sports like these are all their cares beguiled,
> The sports of children satisfy the child."—*Goldsmith.*

Every child, where it be practicable, should have a small plot of ground to cultivate, that he may dig and delve in, and make dirt-pies if he choose. Children now-a-days, unfortunately, are not allowed to soil their hands and their fine clothes. For my own part, I dislike such model children ; let a child be natural—let him,. as far as is possible, choose his own sports. Do not be always interfering with his pursuits, and be finding fault with him. Remember, what may be amusing to you may be distasteful to him. I do not, of course, mean but that you should constantly have a watchful eye over him ; yet do not let him see that he is under restraint or surveillance ; if you do, you will never discover his true character and inclinations. Not only so, but do not dim the bright sunshine of his early life by constantly check-ing and thwarting him. Tupper beautifully says—

" And check not a child in his merriment,—
Should not his morning be sunny ? "

When, therefore, he is either in the nursery or in the play-ground, let him shout and riot and romp about as much as he please. His lungs and his muscles want developing, and his nerves require strengthening ; and how can such be accomplished unless you allow them to be developed and strengthened by natural means ?

The nursery is a child's own domain ; it is his castle, and he should be Lord Paramount therein. If he choose to blow a whistle, or to spring a rattle, or to make any other hideous noise, which to him is sweet music, he should be allowed, without let or hindrance, to do so. If any members of the family have weak nerves, let them keep at a respectful distance.

A child who never gets into mischief must be either sly, or delicate, or idiotic ; indeed, the system of many persons, in bringing up children, is likely to make them either the one or the other. The present plan of train-ing children is nearly all work (books), and very little play. Play, and plenty of it, is necessary to the very existence of a child.

A boy not partial to mischief, innocent mischief, and

play, is unnatural ; he is a man before his time, he is a
nuisance, he is disagreeable to himself and to every one
around. He is generally a sneak, and a little humbug.

Girls, at the present time, are made clever simpletons ;
their brains are worked with useless knowledge, which
totally unfits them for every-day duties. Their muscles
are allowed to be idle, which makes them limp and
flabby. The want of proper exercise ruins the complexion, and their faces become of the colour of a tallow
candle ! And precious wives and mothers they make
when they do grow up ! Grow up, did I say ? They
grow all manner of ways, and are as crooked as crooked
sticks !

What an unnatural thing it is to confine a child
several hours a day to his lessons ; why, you might as
well put a colt in harness, and make him work for his
living ! A child is made for play ; his roguish little
eye, his lithe figure, his antics, and his drollery, all point
out that he is cut out for play—that it is as necessary to
his existence as the food he eats, and as the air he
breathes !

A child ought not to be allowed to have playthings
with which he can injure either himself or others, such
as toy-swords, toy-cannons, toy-paint-boxes, knives, bows
and arrows, hammers, chisels, saws, &c. He will not
only be likely to injure himself and others, but will
make sad havoc on furniture, house, and other property.
Fun, frolic, and play ought, in all innocent ways, to be
encouraged ; but wilful mischief and dangerous games
ought, by every means, to be discountenanced. This
advice is frequently much needed, as children prefer to
have and delight in dangerous toys, and often coax and
persuade weak and indulgent mothers to gratify their
wishes.

Painted toys are, many of them, highly dangerous,
those painted *green* especially, as the colour generally
consists of Scheele's green—arsenite of copper.

Children's paint-boxes are very dangerous toys for a
child to play with : many of the paints are poisonous,

containing arsenic, lead, gamboge, &c.; and a child, when painting, is apt to put the brush into his mouth, to absorb the superabundant fluid. Of all the colours, the *green* paint is the most dangerous, as it is frequently composed of arsenite of copper—arsenic and copper—two deadly poisons.

There are some paint-boxes warranted not to contain a particle of poison of any kind : these ought, for a child, to be chosen by a mother.

But, remember, although he ought not to be allowed to have poison paint-boxes and poison-painted toys, *he must have an abundance of toys*, such as the white-wood toys—brewers' drays, millers' waggons, boxes of wooden bricks, &c. The Noah's Ark is one of the most amusing and instructive toys for a child. " Those fashioned out of brown, unpainted pine-wood by the clever carvers of Nuremberg or the Black Forest are the best, I think, not only because they are the most spirited, but because they will survive a good deal of knocking about, and can be sucked with impunity. From the first dawn of recollection, children are thus familiarised with the forms of natural objects, and may be well up in natural history before they have mastered the A B C."*

Parents often make Sunday a day of gloom : to this I much object. Of all the days in the week, Sunday should be the most cheerful and pleasant. It is considered by our Church a festival ; and a glorious festival it ought to be made, and one on which our Heavenly Father wishes to see all His children happy and full of innocent joy ! Let Sunday, then, be made a cheerful, joyous, innocently happy day, and not, as it frequently is, the most miserable and dismal in the week. It is my firm conviction that many men have been made irreligious by the ridiculously strict and dismal way they were compelled, as children, to spend their Sundays. You can no more make a child religious by gloomy

---

* From an excellent article *About Toys*, by J. Hamilton Fyfe in *Good Words* for December 1862.

asceticism, than you can make people good by Act of Parliament.

One of the great follies of the present age is, children's parties, where they are allowed to be dressed up like grown-up women, stuck out in petticoats, and encouraged to eat rich cake and pastry, and to drink wine, and to sit up late at night ! There is something disgusting and demoralising in all this. Their pure minds are blighted by it. Do not let me be misunderstood : there is not the least objection, but, on the contrary, great advantage, for friends' children to meet friends' children ; but then let them be treated as children, and not as men and women !

180. *Do you approve of public play-grounds for children ?*

It would be well, in every village, and in the outskirts of every town, if a large plot of ground were set apart for children to play in, and to go through regular gymnastic exercises. Play is absolutely necessary to a child's very existence, as much as food and sleep ; but in many parts of England where is he to have it ? Playgrounds and play are the best schools we have ; they teach a great deal not taught elsewhere ; they give lessons in health, which is the grandest wealth that can be bestowed—" for health is wealth : " they prepare the soil for the future schoolmaster ; they clear the brain, and thus the intellect ; they strengthen the muscles ; they make the blood course merrily through the arteries ; they bestow healthy food for the lungs ; they give an appetite ; they make a child, in due time, become every inch a man ! Play-grounds and play are one of the finest institutions we possess. What would our large public schools be without their play and cricket grounds ? They would be shorn of half their splendour and their usefulness !

There is so much talk now-a-days about *useful* knowledge, that the importance of play and play-grounds is likely to be forgotten. I cannot help thinking, however, that a better state of things is dawning. " It seems to be found out that in our zeal for useful knowledge, that knowledge is found to be not the least useful which treat boys as active, stirring, aspiring, and ready."*

EDUCATION.

181. *Do you approve of infant schools ?*

I do, if the arrangements be such that health is preferred before learning.† Let children be only confined for three or four hours a day, and let what little they learn be taught as an amusement rather than as a labour. A play-ground ought to be attached to an infant school; where, in fine weather, for every half-hour they spend in-doors, they should spend one in the open air ; and, in wet weather, they ought to have, in lieu of the play-ground, a large room to romp, and shout, and riot in. To develop the different organs, muscles, and other parts of the body, children require fresh air, a free use of their lungs, active exercise, and their bodies to be thrown into all manner of attitudes. Let a child mope in a corner, and he will become stupid and sickly. The march of intellect, as it is called, or rather the double quick march of intellect, as it should be called, has stolen a march upon health. Only allow the march of intellect and the march of health to take equal strides, and then we shall have " *mens sana in corpore sano* " (a sound mind in a sound body).

In the education of a young child, it is better to instruct him by illustration, by pictures, and by encourag-

---

* *The Saturday Review*, December 13, 1862.

† " According to Aristotle, more care should be taken of the body than of the mind for the first seven years ; strict attention to diet be enforced, &c. . . . . . The eye and ear of the child should be most watchfully and /severely guarded against contamination of every kind, and unrestrained communication with servants be strictly prevented. Even his amusements should be under due regulation, and rendered as interesting and intellectual as possible."—The Rev John Williams, in his *Life and Actions of Alexander the Great*

ing observation on things around and about him, than
by books.   It is surprising how much, without endangering his health, may be taught in this way.   In educating
your child, be careful to instil and to form good habits
—they will then stick to him for life.

Children at the present day are too highly educated—
·their brains are over-taxed, and thus weakened.   The
consequence is; that as they grow up to manhood, if they
grow up at all, they become fools !   *Children* are now
taught what formerly *youths* were taught.   The chord of
a child's life is ofttimes snapped asunder in consequence
of over education :—

"Screw not the cord too sharply, lest it snap."—*Tennyson.*

You should treat a child as you would a young colt.
Think only at first of strengthening his body.   Let him
have a perfectly free, happy life, plenty of food to eat,
abundance of air to breathe, and no work to do ; there is
plenty of time to think of his learning—of giving him
brain work.   It will come sadly too soon; but do not
make him old before his time.

182. *At what age do you advise my child to begin his
course of education—to have his regular lessons ?*

In the name of the prophet,—Figs !   Fiddlesticks !
about courses of education and regular lessons for a child !
You may as well ask me when he, a child, is to begin
Hebrew, the Sanscrit, and Mathematics !   Let him have
a course of education in play ; let him go through regular
lessons in foot-ball, bandy, playing at tic, hares and
hounds, and such like excellent and really useful and
health-giving lessons.   Begin his lessons !   Begin brain
work, and make an idiot of him !   Oh ! for shame, ye
mothers !   You who pretend to love your children so
much, and to tax, otherwise to injure, irreparably to injure their brains, and thus their intellects and their
health, and to shorten their very days.   And all for
what ?   To make prodigies of them !   Forsooth ! to
make fools of them in the end.

183. *Well, then, as you have such a great objection to*

a child commencing his education early in life, at what age may he, with safety, commence his lessons? and which do you prefer—home or school education?

Home is far preferable to a school education. He is, if at home, under your own *immediate* observation, and is not liable to be contaminated by naughty children; for, in every school, there is necessarily a great mixture of the good and of the bad; and a child, unfortunately, is more likely to be led by the bad than by the good. Moreover, if he be educated at home, the mother can see that his brain is not over-worked. At school the brain is apt to be over-worked, and the stomach and the muscles to be under-worked.

Remember, as above stated, *the brain must have but very little work until the child be seven years old:* impress this advice upon your memory, and let no foolish ambition to make your child a clever child allow you, for one moment, to swerve from this advice.

Build up a strong, healthy body, and in due time the brain will bear a *moderate* amount of intellectual labour.

As I have given *you* so much advice, permit me, for one moment, to address a word to the father of your child:—

Let me advise you, then, Mr *Paterfamilias*, to be careful how you converse, what language you use, while in the company of your child. Bear in mind, a child is very observant, and thinks much, weighs well, and seldom forgets all you say and all you do! Let no hasty word, then, and more especially no oath, or no impious language, ever pass your lips, if your child be within hearing. It is, of course, at all times wicked to swear; but it is heinously and unpardonably sinful to swear in the presence of your child! "Childhood is like a mirror, catching and reflecting images. One impious or profane thought, uttered by a parent's lip, may operate upon the young heart like a careless spray of water thrown upon polished steel, staining it with rust, which no after scouring can efface."

Never talk secrets before a child—"little pitchers have long ears;" if you do, and he disclose your secrets—as most likely he will—and thus make mischief, it will be cruel to scold him; you will, for your imprudence, have yourself only to blame. Be most careful, then, in the presence of your child, of what you say, and of whom you speak. This advice, if followed, might save a great deal of annoyance and vexation.

184. *Are you an advocate for a child being taught singing?*

I am: I consider singing a part of his education. Singing expands the walls of his chest, strengthens and invigorates his lungs, gives sweetness to his voice, improves his pronunciation, and is a great pleasure and amusement to him.

SLEEP.

185. *Do you approve of a child sleeping on a* FEATHER *bed?*

A *feather* bed enervates his body, and, if he be so predisposed, causes rickets, and makes him crooked. A horse-hair mattress is the best for a child to lie on. The pillow, too, should be made of horse-hair. A *feather* pillow often causes the head to be bathed in perspiration, thus· enervating the child, and making him liable to catch cold. If he be at all rickety, if he be weak in the neck, if he be inclined to stoop, or if he be at all crooked, let him, by all means, lie without a pillow.

186. *Do you recommend a child, in the middle of the day, to be put to sleep?*

Let him be put on his mattress *awake*, that he may sleep for a couple of hours before dinner, then he will rise both refreshed and strengthened for the remainder of the day. I said, let him be put down *awake*. He might, for the first few times, cry, but, by perseverance, he will without any difficulty fall to sleep. The practice of sleeping before dinner ought to be continued until he be three years old, and, if he can be prevailed upon, even longer. For if he do not have sleep in the middle of the day, he will all the afternoon and the evening be cross; and when he does go to bed, he will probably be too tired to sleep, or his nerves having been exhausted by the long wakefulness, he will fall into a troubled, broken slumber, and not into that sweet, soft, gentle repose, so characteristic of healthy, happy childhood!

187. *At what hour ought a child to be put to bed in the evening?*

At six in the winter, and at seven o'clock in the summer. *Regularity* ought to be observed, *as regularity is very conducive to health.* It is a reprehensible practice to keep a child up until nine or ten o'clock at night. If this be done, he will, before his time, become old, and the seeds of disease will be sown.

As soon as he can run, let him be encouraged, for half an hour before he goes to bed, to race either about the hall, or the landing, or a large room, which will be the best means of warming his feet, of preventing chilblains, and of making him sleep soundly.

188. *Have you any directions to give me as to the placing of my child in his bed?*

If a child lie alone, place him fairly on his side in the middle of the bed; if it be winter time, see that his arms and hands be covered with the bed-clothes; if it be summer, his hands might be allowed to be outside the clothes. In putting him down to sleep, you should ascertain that his face be not covered with the bed-clothes; if it be, he will be poisoned with his own breath—the breath constantly giving off carbonic acid gas; which gas must, if his face be smothered in the clothes, be breathed—carbonic acid gas being highly poisonous.

You can readily prove the existence of carbonic acid gas in the breathing, by simply breathing into a little lime-water; after breathing for a few seconds into it, a white film will form on the top; the carbonic acid gas from the breath unites with the lime of the lime-water, and the product of the white film is carbonate of lime.

189. *Do you advise a bedroom to be darkened at night?*

Certainly : a child sleeps sounder and sweeter in a dark than in a light room. There is nothing better for the purpose of darkening a bedroom, than Venetian blinds. Remember, then, a well-ventilated, but a darkened, chamber at night. The cot or the crib ought not to face the window, " as the light is best behind."[*]

190. *Which is the best position for a child when sleeping—on his back, or on his side ?*

His side : he ought to be accustomed to change about, on the right side one night, on the left another ; and occasionally, for a change, he should lie on his back. By adopting this plan, you will not only improve his figure, but likewise his health. Lying, night after night, in one position, would be likely to make him crooked.

191. *Do you advise, in the winter time, that there should be a fire in the night nursery ?*

Certainly not, unless the weather be intensely cold. I dislike fires in bedrooms, especially for children ; they are very enervating, and make a child liable to catch cold. Cold weather is very bracing, particularly at night. " Generally speaking," says the *Siècle*, " during winter, ·partments are too much heated. The temperature in them ought not to exceed 16° Centigrade (59° Fahrenheit) ; and even in periods of great cold scientific men declare that 12° or 14° had better not be exceeded. In the wards of hospitals, and in the chambers of the sick, care is taken not to have greater heat than 15°. Clerks in offices, and other persons of sedentary occupations, when rooms in which they sit are too much heated, are liable to cerebral [brain] congestion and to pulmonary [lung] complaints. In bedrooms, and particularly those of children, the temperature ought to be maintained rather low ; it is even prudent only rarely to make fires in them, especially during the night."

If " a cold stable make a healthy horse," I am quite sure that a moderately cold and well-ventilated bedroom helps to make a healthy child. But. still. in the winter

---

[*] Sir Charles Locock in a *Letter* to the Author.

time, if the weather be biting cold, a *little* fire in the bedroom grate is desirable. In bringing up children, we must never run into extremes—the coddling system and the hardening system are both to be deprecated ; the coddling system will make the strong child weakly, while the hardening system will probably kill a delicate one. ↓

A child's bed ought, of course, to be comfortably clothed with blankets—I say blankets, as they are much superior to coverlids ; the perspiration will more readily pass through a blanket than a coverlid. A *thick* coverlid ought never to be used ; there is nothing better, for a child's bed, than the old-fashioned patchwork coverlid, as the perspiration will easily escape through it.

192. *Should a child be washed and dressed* AS SOON AS HE AWAKE *in the morning ?*

He ought, if he awake in anything like reasonable time ; for if he doze after he be once awake, such slumber does him more harm than good. He should be up every morning as soon as it is light. If, as a child, he be taught to rise early, it will make him an early riser for life, and will tend greatly to prolong both his existence and his happiness.

Never awake a child from his sleep to dress him, to give him medicine, or for any other purpose ; *let him always sleep as long as he can ;* but the moment he awakes let him be held out, and then let him be washed and dressed, and do not wait, as many a silly nurse does, until he have wet his bed, until his blood be chilled, and until he be cross, miserable, and uncomfortable ! How many babes are made ill by such foolish practices !

The moment he leaves his bed, turn back to the fullest extent the clothes, in order that they may be thoroughly ventilated and sweetened. They ought to be exposed to the air for at least an hour before the bed be made. As soon as he leaves his room, be it winter or summer, throw open the windows.

193. *Ought a child to lie alone ?*

He should, after he is weaned. He will rest more comfortably, and his sleep will be more refreshing.

194. *Supposing a child should not sleep well, what ought to be done ?   Would you give him a dose of composing medicine ?*

Certainly not.   Try the effects of exercise.   Exercise in the open air is the best composing medicine in the world.   Let the little fellow be well tired out, and there will be little fear of his not sleeping.

195. *Have you any further observations to make on the subject of sleep ?*

Send a child joyful to bed.   Do not, if you can possibly help it, let him go to bed crying.   Let the last impressions he has at night be of his happy home, and of his loving father and mother, and let his last thoughts be those of joy and gladness.   He will sleep all the sounder if he be sent to bed in such a frame of mind, and he will be more refreshed and nourished in the morning by his sleep.

196. *What are the usual causes of a child walking in his sleep, and what measures, during such times, ought to be adopted to prevent his injuring himself ?*

A disordered stomach, in a child of nervous temperament, or worms, are usually the causes.   The means to be adopted to prevent his throwing himself out of the window, are to have bars to his chamber casement, and if that be not practicable, to have either nails or screws driven into the window sash to allow the window to open only for a sufficient space for ventilation, and to have a screw window fastening, in order that he cannot, without difficulty, open the window; to have a trusty person to sleep in his room, who should have directions given not to rouse him from his sleep, but to gently lead him back to his bed, which may frequently be done without awaking him; and to consult a medical man, who will adopt means to destroy the worms, to put his stomach into order, to brace his nerves, and to strengthen his general system.   A trip to the coast and sea-bathing, in such a case, is often of great service.

SECOND DENTITION.

197. *When does a child commence to cut his* SECOND *set of teeth ?*

Generally at seven years old. He *begins* to *cut* them at about that time; but it should be borne in mind (so wonderful are the works of God) that the *second* crop of teeth, *in embryo*, is actually bred and formed from the very commencement, of his life, *under* the first tier of teeth, but which remain in abeyance for years, and do not come into play until the *first* teeth, having done their duty, loosen and fall out, and thus make room for *the* more numerous, larger, stronger, and more permanent teeth, which latter have to last for the remainder of his existence. The *first* set is sometimes cut with a great deal of difficulty, and produces various diseases; the *second*, or permanent teeth, come easily, and are unaccompanied with any disorder. The following is the process:—One after another of the *first* set gradually loosen, and either drop out, or with little pain are readily pulled out; under these, the *second*—the permanent— teeth make their appearance, and fill up the vacant spaces. The fang of the tooth that has dropped out is nearly all absorbed or eaten away, leaving little more than the crown. The *first* set consists of twenty; the *second* (including the wise-teeth, which are not, generally cut until after the age of twenty-seven) consists of thirty-two.

I would recommend you to pay particular attention to the teeth of your children; for, besides their being ornamental, their regularity and soundness are of great importance to the present as well as to the future health of your offspring. If there be any irregularity in the appearance of the *second* set, lose no time in consulting an experienced and respectable dentist.

ON DISEASE, ETC.

198. *Do you think it important that I should be made acquainted with the symptoms of the* SERIOUS *diseases of children.*

Certainly. I am not advocating the doctrine of a mother *treating serious* diseases; far from it, it is not her province, except in certain cases of extreme urgency, where a medical man cannot be procured, and where delay might be death; but I do insist upon the necessity of her knowing the *symptoms* of disease. My belief is, that if parents were better informed on such subjects, many children's lives might be saved, much suffering averted, and sorrow spared. The fact is, the knowledge of the symptoms of disease is, to a mother, almost a sealed book. If she were better acquainted with these matters, how much more useful would she be in a sick-room, and how much more readily would she enter into the plans and views of the medical man! By her knowledge of the symptoms, and by having his advice in time, she would nip disease in the bud, and the fight might end in favour of life, for "sickness is just a fight between life and death."—*Geo. M'Donald.*

It is really lamentable to contemplate the amount of ignorance that still exists among mothers in all that appertains to the diseases of children; although, fortunately, they are beginning to see and to feel the importance of gaining instruction on such subjects; but the light is only dawning. A writer of the *Medical Times and Gazette* makes the following remarks, which somewhat bear on the subject in question. He observes—"In spite of the knowledge and clear views possessed by the profession on all that concerns the management of children, no fact is more palpable than that the most grievous ignorance and incompetency prevail respecting it among the public. We want some means of making popular the knowledge which is now almost restricted to medical men, or, at most, to the well-educated classes."

In the earlier editions of this work I did not give the *treatment* of any serious diseases, however urgent. In the eight last editions, I have been induced, for reasons I will presently state, to give the *treatment* of some of the more urgent *serious* diseases, when a medical man cannot instantly be procured, and where delay might be death.

Sir CHARLES LOCOCK, who has taken a kind interest in this little work, has given me valid reasons why a mother should be so enlightened. The following extracts are from a letter which I received from Sir CHARLES on the subject, and which he has courteously allowed me to publish. He says,—" As an old physician of some experience in complaints of infants and children, I may perhaps be allowed to suggest that in a future edition you should add a few words on the actual treatment of some of the more urgent infantile diseases. It is very right to caution parents against superseding the doctor, and attempting to manage serious illness themselves ; but your advice, with very small exceptions, always being, ' to lose no time in sending for a medical man,' much valuable and often irremediable time may be lost *when a medical man is not to be had.* Take, for instance, a case of croup : there are no directions given at all, except to send for a medical man, and always to keep medicines in the house which he may have directed. But how can this apply to a first attack ? You state that a first attack is generally the worst. But why is it so ? Simply because it often occurs when the parents do not recognise it, and it is allowed to get a worse point than in subsequent attacks, when they are thoroughly alive to it. As the very best remedy, and often the only essential one, if given early, is a full emetic, surely it is better that you should give some directions as to this in a future edition, and I can speak from my own experience when I say that an emetic, *given in time*, and repeated to free vomiting, will cut short *any* case of croup. In nine cases out of ten the attack takes place in the evening or early night, and when vomiting is effected the dinner of that day is brought up nearly undigested, and the severity of the symptoms at once cut short. Whenever any remedy is valuable, the more by its being administered *in time*, it is surely wiser to give directions as to its use, although, as a general rule, it is much better to advise the sending for medical advice."

The above reasons, coming from such a learned and experienced physician as Sir Charles Locock, are conclusive, and have decided me to comply with his advice, to enlighten a mother on the *treatment* of some of the more urgent diseases of infants and of children. In a subsequent letter addressed to myself, Sir Charles has given me the names of those *urgent* diseases, which he considers may be treated by a mother, " where a medical man cannot be procured quickly, or not at all;" they are—Croup; Inflammation of the Lungs; Diphtheria; Dysentery; Diarrhœa; Hooping-cough, in its various stages; and Shivering Fit. Sir Charles sums up his letter to me by saying,—" Such a book ought to be made as complete as possible, and the objections to medical treatment being so explained as to induce mothers to try to avoid medical men is not so serious as that of leaving them without any guide in those instances where every delay is dangerous, and yet where medical assistance is not to be obtained or not to be had quickly."

In addition to the above, I shall give you the *treatment* of Bronchitis, Measles, and Scarlet Fever. Bronchitis is one of the most common diseases incidental to childhood, and, with judicious treatment, is, in the absence of the medical man, readily managed by a sensible mother. Measles is very submissive to treatment. Scarlet Fever, *if it be not malignant*, and, *if it be not complicated with diphtheric-croup*, and if certain rules be strictly followed, is also equally amenable to treatment.

I have been fortunate in treating Scarlet Fever, and I therefore think it desirable to enter fully into the *treatment* of a disease which is looked upon by many parents, and, according to the usual mode of treatment, with just cause, with great consternation and dread. By giving my plan of treatment, fully and simply, and without the slightest reservation, I am fully persuaded, through God's blessing, that I may be the humble means of saving the lives of numbers of children.

The diseases that might be treated by a mother, in the absence of a medical man, will form the subject of future Conversations.

I think it right to premise that in all the prescriptions for a child I have for the use of a mother given, I have endeavoured to make them as simple as possible, and have, whenever practicable, avoided to recommend powerful drugs. Complicated prescriptions and powerful medicines ought, as a rule, to be seldom given; and when they are, should only be administered by a judicious medical man; a child requiring much more care and gentleness in his treatment than an adult; indeed, I often think it would be better to leave a child to nature rather than to give him powerful and large doses of medicines. A remedy—calomel, for instance—has frequently done more mischief than the disease itself; and the misfortune of it is, the mischief from that drug has oftentimes been permanent, while the complaint might, if left alone, have only been temporary.

199. *At what age does Water on the Brain usually occur, and how is a mother to know that her child is about to labour under that disease ?*

Water on the brain is, as a rule, a disease of childhood : after a child is seven years old it is comparatively rare. It more frequently attacks delicate children—children who have been dry-nursed (especially if they have been improperly fed), or who have been suckled too long, or who have had consumptive mothers, or who have suffered severely from teething, or who are naturally of a feeble constitution. Water on the brain sometimes follows an attack of inflammation of the lungs, more especially if depressing measures (such as excessive leeching and the administration of emetic tartar) have been adopted. It occasionally follows in the train of contagious eruptive diseases, such as either small-pox or scarlatina. We may divide the symptoms of water on the brain into two stages. The first—the premonitory stage—which lasts four or five days, in which medical aid might be of great avail: the second—the stage of drowsiness and of coma—which usually ends in death.

I shall dwell on the first—the premonitory stage—in order that a mother may see the importance without loss of time of calling in a medical man :—

If her child be feverish and irritable, if his stomach be disordered, if he have urgent vomitings, if he have a foul breath, if his appetite be capricious and bad if his nights be disturbed (screaming out in his sleep), if his bowels be disordered, more especially if they be con- stipated, if he be more than usually excited, if his eye gleam with unusual brilliancy, if his tongue run faster than it is wont, if his cheek be flushed and his head be hot, and if he be constantly putting his hand to his head ; there is cause for suspicion. If to these symptoms be added, a more than usual carelessness in tumbling about, in hitching his foot in the carpet, or in dragging one foot after the other ; if, too, he has complained of darting, shooting, lancinating pains in his head, it may then be known that the *first* stage of inflammation (the forerunner of water on the brain) either has taken, or is about taking place. Remember no time ought to be lost in obtaining medical aid ; for the *commencement* of the disease is the golden opportunity, when life might pro- bably be saved.

200. *At what age, and in what neighbourhood, is a child most liable to croup, and when is a mother to know that it is about to take place ?*

It is unusual for a child until he be twelve months old to have croup : but, from that time until the age of two years, he is more liable to it than at any other period. The liability after two years, gradually, until he be ten years old, lessens, after which time it is rare.

A child is more liable to croup in a low and damp, than in a high and dry neighbourhood ; indeed, in some situations, croup is almost an unknown disease ; while in others it is only too well understood. Croup is more likely to prevail when the wind is either easterly or north-easterly.

There is no disease that requires more prompt treatment than croup, and none that creeps on more insidiously. The child at first seems to be labouring under a slight cold, and is troubled with a little *dry* cough ; he is hot and fretful, and *hoarse* when he cries. Hoarseness is one of the earliest symptoms of croup ; and it should be borne in mind that a young child, unless he be going to have croup, is seldom hoarse ; if, therefore, your child be hoarse, he should be carefully watched, in order that, as soon as croup be detected, not a moment be lost in applying the proper remedies.

His voice at length becomes gruff, he breathes as though it were through muslin, and the cough becomes crowing. These three symptoms prove that the disease is now fully formed. These latter symptoms sometimes come on without any previous warning, the little fellow going to bed apparently quite well, until the mother is awakened, perplexed and frightened, in the middle of the night, by finding him labouring under the characteristic cough and the other symptoms of croup. If she delay either to send for assistance, *or if proper medicines be not instantly given*, in a few hours it will probably be of no avail, and in a day or two the little sufferer will be a corpse !

When once a child has had croup the after attacks are generally milder. If he has once had an attack of croup, I should advise you always to have in the house medicine—a 4 oz. bottle of Ipecacuanha Wine, to fly to at a moment's notice,* but never omit, where practicable, in a case of croup, whether the case be severe or mild, to send *immediately* for medical aid. There is no disease in which time is more precious than in croup, and where the delay of an hour may decide either for life or for death.

201. *But suppose a medical man is not* IMMEDIATELY

---

* In case of a sudden attack of croup, *instantly* give a teaspoonful of Ipecacuanha Wine, and repeat it every five minutes until free vomiting be excited.

*to be procured, what then am I to do ? more especially,
as you say, that delay might be death ?*

What to do.—I never, in my life, lost a child with
croup—with catarrhal-croup—where I was called in at
the *commencement* of the disease, and where my plans
were carried out to the very letter. Let me begin by
saying, look well to the goodness and purity of the
medicine, for the life of your child may depend upon the
medicine being genuine. What medicine ? *Ipecacuanha
Wine !* At the earliest dawn of the disease give a tea-
spoonful of Ipecacuanha Wine every five ₁minutes, until
free vomiting be excited. In croup, then, before he be
safe free vomiting *must* be established, and that without
loss of time. If, after the expiration of an hour, the
Ipecacuanha Wine (having given during that hour one
or two tea-spoonfuls of it every five minutes) be not
sufficiently powerful for the purpose—although it
generally is so—(*if the Ipecacuanha Wine be good*)—
then let the following mixture be substituted :—

Take of—Powdered Ipecacuanha, one scruple ;
      Wine of Ipecacuanha, one ounce and a half :
Make a Mixture. One or two tea-spoonfuls to be given every
five minutes, until free vomiting be excited, first well shaking
the bottle.

After the vomiting, place the child for a quarter of an
hour in a warm bath.* When out of the bath give him
small doses of Ipecacuanha Wine every two or three
hours. The following is a palatable form for the mixture :—

Take of—Wine of Ipecacuanha, three drachms ;
      Simple syrup, three drachms ;
      Water, six drachms :
Make a Mixture. A tea-spoonful to be taken every two or
three hours.

But remember the emetic which is given at *first* is *pure
Ipecacuanha Wine, without a drop of either water or of
syrup.*

---

* See " Warm Baths "—directions and precautions to be observed.

A large sponge dipped out of very hot water, and applied to the throat, and frequently renewed, oftentimes affords great relief in croup, and ought during the time the emetic is being administered in all cases to be adopted.

If it be a *severe* case of croup, and does not in the course of two hours yield to the free exhibition of the Ipecacuanha Emetic, apply a narrow strip of *Smith's Tela Vesicatoria* to the throat, prepared in the same way as for a case of inflammation of the lungs (see the Conversation on the *treatment* of inflammation of the lungs). With this only difference, let it be a narrower strip, only one-half the width there recommended, and apply it to the throat instead of to the chest. If a child has a very short, fat neck, there may not be room for the *Tela*, then you ought to apply it to the *upper* part of the chest—just under the collar-bones.

Let it be understood, that the *Tela Vesicatoria* is not a severe remedy, that the *Tela* produces very little pain —not nearly so much as the application of leeches; although, in its action, it is much more beneficial, and is not nearly so weakening to the system.

Keep the child from all stimulants ; let him live on a low diet, such as milk and water, toast and water, arrow-root, &c. ; and let the room be, if practicable, at a temperate heat—60° Fahrenheit, and be well ventilated.

So you see that the *treatment* of croup is very simple, and that the plan might be carried out by an intelligent mother. Notwithstanding which, it is your duty, where practicable, to send, at the very *onset* of the disease, for a medical man.

Let me again reiterate that, if your child is to be saved, the *Ipecacuanha Wine must be genuine and good.* This can only be effected by having the medicine from a highly respectable chemist. Again, if ever your child has had croup, let me again urge you *always* to have in the house a 4 oz. bottle of Ipecacuanha Wine, that you may resort to at a moment's notice, in case there be the slightest return of the disease.

Ipecacuanha Wine, unfortunately, is not a medicine that keeps well; therefore, every three or four months a fresh bottle ought to be procured, either from a medical man or from a chemist. As long as the Ipecacuanha Wine remains *clear*, it is good; but as soon as it becomes *turbid*, it is bad, and ought to be replaced by a fresh supply. An intelligent correspondent of mine makes the following valuable remarks on the preservation of Ipecacuanha Wine :—" Now, I know that there are some medicines and chemical preparations which, though they spoil rapidly when at all exposed to the air, yet will keep perfectly good for an indefinite time if hermetically sealed up in a *perfectly full* bottle. If so, would it not be a valuable suggestion if the Apothecaries' Hall, or some other London firm of *undoubted* reliability, would put up 1 oz. phials of Ipecacuanha Wine of guaranteed purity, sealed up so as to keep good so long as unopened, and sent out in sealed packages, with the guarantee of their name. By their keeping a few such ounce bottles in an unopened state in one's house, one might rely in being ready for any emergency. If you think this suggestion worth notice, and could induce some first-rate house to carry it out, and mention the fact in a subsequent edition of your book, you would, I think, be adding another most valuable item to an already invaluable book."

The above suggestion of preserving Ipecacuanha Wine in ounce bottles, quite full, and hermetically sealed, is a very good one. The best way of hermetically sealing the bottle would be, to cut the cork level with the lip of the bottle, and to cover the cork with sealing-wax, in the same manner wine merchants serve some kinds of their wines, and then to lay the bottles on their sides in sawdust in the cellar. I have no doubt, if such a plan were adopted, the Ipecacuanha Wine would for a length of time keep good. Of course, if the Wine of Ipecacuanha be procured from the Apothecaries' Hall Company, London (as suggested by my correspondent), there can be no question as to the genuineness of the article.

What NOT *to do.*—Do not give emetic tartar; do not apply leeches; do not keep the room very warm; do not give stimulants; do not omit to have always in the house either a 4 oz. bottle, or three or four 1 oz. bottles, of Ipecacuanha Wine.

202. *I have heard Child-crowing mentioned as a formidable disease; would you describe the symptoms?*

Child-crowing, or spasm of the glottis, or *spurious croup*, as it is sometimes called, is occasionally mistaken for *genuine croup*. It is a more frequent disorder than the latter, and requires a different plan of treatment. Child-crowing is a disease that invariably occurs only during dentition, and is *most perilous;* indeed, painful dentition is *the* cause—*the* only cause—of child-crowing. But, if a child labouring under it can fortunately escape suffocation until he have cut the whole of his first set of teeth—twenty—he is then safe.

Child-crowing comes on in paroxysms. The breathing during the intervals is quite natural—indeed, the child appears perfectly well; hence, the dangerous nature of the disease is either overlooked, or is lightly thought of, until perhaps a paroxysm worse than common takes place, and the little patient dies of suffocation, overwhelming the mother with terror, with confusion, and dismay.

The *symptoms* in a paroxysm of child-crowing are as follows:—The child suddenly loses and fights for his breath, and in doing so, makes a noise very much like that of crowing; hence the name child-crowing. The face during the paroxysm becomes bluish or livid. In a favourable case, after either a few seconds, or even, in some instances, a minute, and a frightful struggle to breathe, he regains his breath, and is, until another paroxysm occurs, perfectly well. In an unfavourable case, the upper part (chink) of the windpipe—the glottis—remains for a minute or two closed, and the child, not being able to breathe, drops a corpse in his nurse's arms! Many children, who are said to have died of fits, have really died of child-crowing.

Child-crowing is very apt to cause convulsions, which complication, of course, adds very much to the danger. Such a complication requires the constant supervision of an experienced and skilful medical man.

I have entered thus rather fully into the subject, as nearly every life might be saved, if a mother knew the nature and the treatment of the complaint, and of the *great necessity during the paroxysm of prompt and proper measures.* For, too frequently, before a medical man has had time to arrive, the child has breathed his last, the parent herself being perfectly ignorant of the necessary treatment; hence the vital importance of the subject, and the paramount necessity of imparting such information, in a *popular* style, in conversations of this kind.

203. *What treatment, then, during a paroxysm of Child-crowing should you advise ?*

The first thing, of course, to be done, is to send *immediately* for a medical man. Have a plentiful supply of cold and of hot water always at hand, ready at a moment's notice for use. The instant the paroxysm is upon the child, plentifully and perseveringly dash *cold* water upon his head and face. Put his feet and legs in *hot* salt, mustard, and water; and, if necessary, place him up to his neck in a hot bath, still dashing water upon his face and head. If he does not quickly come round, sharply smack his back and buttocks.

In every severe paroxysm of child-crowing, put your fore-finger down the throat of the child, and pull his tongue forward. This plan of pulling the tongue forward opens the epiglottis (the lid of the glottis), and thus admits air (which is so sorely needed) into the glottis and into the lungs, and thus staves off impending suffocation. If this plan were generally known and adopted, many precious lives might be saved.*

---

* An intelligent correspondent first drew my attention to the efficacy of pulling forward the tongue in every severe paroxysm of child-crowing.

There is nothing more frightfully agonising to a mother's feelings than to see her child strangled,—as it were,—before her eyes, by a paroxysm of child-crowing.

As soon as a medical man arrives, he will lose no time in thoroughly lancing the gums, and in applying other appropriate remedies.

Great care and attention ought, during the intervals, to be paid to his diet. If the child be breathing a smoky, close atmosphere, he should be immediately removed to a pure one. In this disease, indeed, there is no remedy equal to a change of air—to a dry, bracing neighbourhood. Change of air, even if it be winter, is the best remedy, either to the coast or to a healthy mountainous district. I am indebted to Mr Roberton of Manchester (who has paid great attention to this disease, and who has written a valuable essay on the subject*) for the knowledge of this fact. Where, in a case of this kind, it is not practicable to send a child *from* home, then let him be sent out of doors the greater part of every day; let him, in point of fact, almost live in the open air. I am quite sure, from an extensive experience, that in this disease, fresh air, and plenty of it, is the best and principal remedy. Cold sponging of the body too is useful.

Mr Roberton, who, at my request, has kindly given me the benefit of his extensive experience in child-crowing, considers that there is no remedy, in this complaint, equal to fresh air—to dry cold winds—that the little patient ought, in fact, nearly to live, during the day, out of doors, whether the wind be in the east or in the north-east, whether it be biting cold or otherwise, provided it be dry and bracing, for "if the air be dry, the colder the better,"—taking care, of course, that he be well wrapped up. Mr Roberton, moreover, advises that the child should be sent away at once from home, either to a bracing sea-side place, such as Blackpool or Fleetwood; or to a mountainous district, such as Buxton.

* See the end of the volume of "Physiology and Diseases of Women,"·&c. Churchill, 1851.

As the subject is so important, let me recapitulate : the gums ought, from time to time, to be well lanced, in order to remove the irritation of painful dentition—painful dentition being the real cause of the disease. Cold sponging should be used twice or thrice daily. The diet should be carefully attended to (see Dietary of Child) ; and everything conducive to health should (as recommended in these Conversations) be observed. But, remember, after all that can be said about the treatment, there is nothing like change of air, of fresh air, of cold, dry pure air, and of plenty of it—the more the little fellow can inhale, during the day, the better it will be for him, it will be far better than any drug contained in the pharmacopœia.

I have dwelt on this subject at some length—it being a most important one—as, if the above advice were more generally known and followed, nearly every child, labouring under this complaint, would be saved ; while now, as coroners' inquests abundantly testify, the disease carries off yearly an immense number of victims.

204. *When is a mother to know that a cough is not a " tooth cough," but one of the symptoms of Inflammation of the lungs ?*

If the child has had a shivering fit ; if his skin be very hot and very dry ; if his lips be parched ; if there be great thirst ; if his cheeks be flushed ; if he be dull and heavy, wishing to be quiet in his cot or crib ; if his appetite be diminished ; if his tongue be furred ; if his mouth be *burning* hot and dry ;* if his urine be scanty and high-coloured, staining the napkin, or the linen ; *if his breathing be short, panting, hurried, and oppressed ; if there be a hard dry cough ; and if his skin be burning hot ;*—then there is no doubt that inflammation of the lungs has taken place.

No time should be lost in sending for medical aid ;

---

* If you put your finger into the mouth of a child labouring under inflammation of the lungs, it is like putting your finger into a hot apple pie, the heat is so great

indeed, the *hot, dry mouth and skin, and short, hurried breathing* would be sufficient cause for your procuring *immediate* assistance. If inflammation of the lungs were properly treated at the *onset*, a child would scarcely ever be lost by that disease. I say this advisedly, for in my own practice, *provided I am called in early, and if my plans are strictly carried out*, I scarcely ever lose a child from inflammation of the lungs.

You may ask—What are your plans? I will tell you, in case *you cannot promptly obtain medical advice*, as delay might be death!

The treatment of Inflammation of the Lungs, what to do.—Keep the child to one room, to his bedroom, and to his bed. Let the chamber be properly ventilated. If the weather be cool, let a small fire be in the grate; otherwise, he is better without a fire. Let him live on low diet, such as weak black tea, milk and water (in equal quantities), and toast and water, thin oatmeal gruel, arrow-root, and such like simple beverages, and give him the following mixture :—

> Take of—Wine of Ipecacuanha, three drachms ;
> Simple Syrup, three drachms;
> Water, six drachms :
> Make a Mixture. A tea-spoonful of the mixture to be taken every four hours.

Be careful that you go to a respectable chemist, in order *that the quality of the Ipecacuanha Wine may be good, as the child's life may depend upon it.*

If the medicine produce sickness, so much the better; continue it regularly until the short, oppressed, and hurried breathing has subsided, and has become natural.

If the attack be very severe, in addition to the above medicine, at once apply a blister, not the common blister, but *Smith's Tela Vesicatoria**—a quarter of a sheet. If the child be a year old, the blister ought to be kept on for three hours, and then a piece of dry, soft

---

* Manufactured by T. & H. Smith, chemists, Edinburgh, and may be procured of Southalls, chemists, Birmingham.

linen rag should be applied for another three hours. **At**
the end of which time—six hours—there will be a
beautiful blister, which must then, with a pair of
scissors, be cut, to let out the water ; and then let the
blister be dressed, night and morning, with simple
cerate spread on lint.

If the little patient be more than one year, say two
years old, let the Tela remain on for five hours, and the
dry linen rag for five hours more, before the blister, as
above recommended, be cut and dressed.

If in a day or two the inflammation still continue
violent, let another Tela Vesicatoria be applied, not over
the old blister, but let a narrow strip of it be applied on
each side of the old blister, and managed in the same
manner as before directed.

I cannot speak too highly of Smith's Tela Vesicatoria.
It has, in my hands, through God's blessing, saved the
lives of scores of children.   It is far, very far, superior
to the old-fashioned blistering plaster.   It seldom, if the
above rules be strictly observed, fails to rise ; it gives
much less pain than the common blister ; when it has
had the desired effect, it readily heals, which cannot
always be said of the common fly-blister, more
especially with children.

My sheet anchors, then, in the inflammation of the
lungs of children are, Ipecacuanha Wine and Smith's
Tela Vesicatoria.   Let the greatest care, as I before
advised, be observed in obtaining the Ipecacuanha Wine
genuine and good.   This can be only depended upon by
having the medicine from a highly respectable chemist.
Ipecacuanha Wine, when genuine and good, is, in many
children's diseases, one of the most valuable of
medincies.

What, in a case of inflammation of the lungs, NOT *to
do.*—Do not, on any account, apply leeches.   They
draw out the life of the child, but not his disease.
Avoid—*emphatically let me say so*—giving emetic tartar.
It is one of the most lowering and death-dealing
medicines that can be administered either to an infant

or to a child ! If you wish to try the effect of it, take a dose yourself, and I am quite sure that you will then never be inclined to poison a child with such an abominable preparation ! In olden times—many, many years ago—I myself gave it in inflammation of the lungs, and lost many children ! Since leaving it off, the recoveries of patients by the Ipecacuanha treatment, combined with the external application of Smith's *Tela Vesicatoria*, have been in many cases marvellous. Avoid · broths and wine, and all stimulants. Do *not* put the child into a warm bath; it only oppresses the already oppressed breathing. Moreover, after he is out of the bath, it causes a larger quantity of blood to rush back to the lungs and to the bronchial tubes, and thus feeds the inflammation. Do not, by a large fire, keep the temperature of the room high. A small fire, in the winter time, encourages ventilation, and in such a case does good. When the little patient is on the mother's or on the nurse's lap, do not burden him either with a *heavy* blanket or with a *thick* shawl. Either a *thin* child's blanket, or a thin *woollen* shawl, in addition to his usual nightgown, is all the clothing necessary.

205. *Is Bronchitis a more frequent disease than In-flammation of the Lungs? Which is the most dangerous? What are the symptoms of Bronchitis?*

Bronchitis is a much more frequent disease than inflammation of the lungs; indeed, it is one of the most common complaints both of infants and of children, while inflammation of the lungs is comparatively a rare disease. Bronchitis is not nearly such a dangerous disease as inflammation of the lungs.

The symptoms.—The child for the first few days labours under symptoms of a heavy cold; he has not his usual spirits. In two or three days, instead of the cold leaving him, it becomes more confirmed; he is now really poorly, fretful, and feverish; his breathing becomes rather hurried and oppressed; his cough is hard and dry, and loud; he wheezes, and if you put your ear to his naked back, between his shoulder blades, you will hear the wheezing more distinctly. If at the breast, he does not suck with his usual avidity; the cough, notwithstanding the breast is a great comfort to him, compels him frequently to loose the nipple; his urine is scanty, and rather high-coloured, staining the napkin, and smelling strongly. He is generally worse at night.

Well, then, remember if the child be feverish, if he have symptoms of a heavy cold, if he have an oppression of breathing, if he wheeze, and if he have a tight, dry, . noisy cough, you may be satisfied that he has an attack of bronchitis.

206. *How can I distinguish between Bronchitis and Inflammation of the Lungs?*

In bronchitis the skin is warm, but moist; in inflammation of the lungs it is hot and dry : in bronchitis the mouth is warmer than usual, but moist; in inflammation of the lungs it is burning hot : in bronchitis the breathing is rather hurried, and attended with wheezing; in inflammation of the lungs it is very short and panting, and is unaccompanied with wheezing, although occasionally a very slight crackling sound might be heard : in bronchitis the cough is long and noisy; in inflammation of the lungs it is short and feeble : in bronchitis the child is cross and fretful; in inflammation of the lungs he is dull and heavy, and his countenance denotes distress.

We have sometimes a combination of bronchitis and of inflammation of the lungs, an attack of the latter following the former. Then the symptoms will be modified, and will partake of the character of the two diseases.

207. *How would you treat a case of Bronchitis?*

If a medical man cannot be procured, I will tell you *What to do :* Confine the child to his bedroom, and if very ill, to his bed. If it be winter time, have a little fire in the grate, but be sure that the temperature of the chamber be not above 60° Fahrenheit, and let the room be properly ventilated, which may be effected by occasionally leaving the door a little ajar.

Let him lie either *outside* the bed or on a sofa ; if he be very ill, *inside* the bed, with a sheet and a blanket only to cover him, but no thick coverlid. If he be allowed to lie on the lap, it only heats him and makes him restless. If he will not lie on the bed, let him rest on a pillow placed on the lap ; the pillow will cause him to lie cooler, and will more comfortably rest his wearied body. If he be at the breast, keep him to it ; let him have no artificial food, unless, if he be thirsty, a little toast and water. If he be weaned, let him have either milk and water, arrow-root made with equal parts of milk and water, toast and water, barley water, or weak black tea, with plenty of new milk in it, &c. ; but, until the inflammation have subsided, neither broth nor beef-tea.

Now, with regard to medicine, the best medicine is Ipecacuanha Wine, given in large doses, so as to produce constant nausea. The Ipecacuanha abates fever, acts on the skin, loosens the cough, and, in point of fact, in the majority of cases, will rapidly effect a cure. I have in a preceding Conversation given you a prescription for the Ipecacuanha Wine Mixture. Let a tea-spoonful of the mixture be taken every four hours.

If in a day or two he be no better, but worse, by all means continue the mixture, whether it produce sickness or otherwise ; and put on the chest a *Tela Vesicatoria*, a quarter of a sheet.

The Ipecacuanha Wine and the Tela Vesicatoria are my sheet-anchors in the bronchitis, both of infants and of children. They rarely, even in very severe cases, fail to effect a cure, provided the Tela Vesicatoria be properly applied, and the Ipecacuanha Wine be genuine and of good quality.

If there be any difficulty in procuring *good* Ipecacuanha Wine, the Ipecacuanha may be given in powder instead of the wine. The following is a pleasant form :—

Take of Powder of Ipecacuanha, twelve grains
,,   White Sugar, thirty-six grains :
Mix well together, and divide into twelve powders. One of the powders to be put dry on the tongue every four hours.

The Ipecacuanha Powder will keep better than the Wine—an important consideration to those living in country places ; nevertheless, if the Wine can be procured fresh and good, I far prefer the Wine to the Powder.

When the bronchitis has disappeared, the diet ought gradually to be improved—rice, sago, tapioca, and light batter-pudding, &c.; and, in a few days, either a little chicken or a mutton chop, mixed with a well-mashed potato and crumb of bread, should be given. But let the improvement in his diet be gradual, or the inflammation might return.

What NOT *to do.*—Do not apply leeches. Do not give either emetic tartar or antimonial wine, which is emetic tartar dissolved in wine. Do not administer either paregoric or syrup of poppies, either of which would stop the cough, and would thus prevent the expulsion of the phlegm. Any fool can stop a cough, but it requires a wise man to rectify the mischief. A cough is an effort of Nature to bring up the phlegm, which would otherwise accumulate, and in the end cause death. Again, therefore, let me urge upon you the immense importance of *not* stopping the cough of a child. The Ipecacuanha Wine will, by loosening the phlegm, loosen the cough, which is the only right way to get rid of a cough. Let what I have now said be impressed deeply upon your memory, as thousands of children in England are annually destroyed by having their coughs stopped. Avoid, until the bronchitis be relieved, giving him broths, and meat, and stimulants of all kinds. For further observations on *what* NOT *to do* in bronchitis, I beg to refer you to a previous Conversation we had on *what* NOT *to do* in inflammation of the lungs. That which is injurious in the one case is equally so in the other.

208. *What are the symptoms of Diphtheria, or, as it is sometimes called, Boulogne Sore-throat ?*

This terrible disease, although by many considered to be a new complaint, is, in point of fact, of very ancient origin. Homer, and Hippocrates, the Father of Physic, have both described it. Diphtheria first appeared in England in the beginning of the year 1857, since which time it has never totally left our shores.

The symptoms.—The little patient, before the disease really shows itself, feels poorly, and is " out of sorts." A shivering fit, though not severe, may generally be noticed. There is heaviness, and slight headache, principally over the eyes. Sometimes, but not always, there is a mild attack of delirium at night. The next day he complains of slight difficulty of swallowing. If old enough, he will complain of constriction about the swallow. On examining the throat, the tonsils will be found to be swollen and redder—more darkly red than usual. Slight specks will be noticed on the tonsils. In a day or two an exudation will cover them, the back of the swallow, the palate, the tongue, and sometimes the inside of the cheeks and of the nostrils. · This exudation of lymph gradually increases until it becomes a regular membrane, which puts on the appearance of leather; hence its name diphtheria. This membrane peels off in pieces; and if the child be old and strong enough he will sometimes spit it up in quantities, the membrane again and again rapidly forming as before. The discharges from the throat are occasionally, but not always, offensive. There is danger of croup from the extension of the membrane into the wind-pipe. The glands about the neck and under the jaw are generally much swollen ; the skin is rather cold and clammy ; the urine is scanty and usually pale; the bowels at first are frequently relaxed. This diarrhœa may, or may not, cease as the disease advances.

The child is now in a perilous condition, and it becomes a battle between his constitution and the disease. If, unfortunately, as is too often the case—diphtheria being more likely to attack the weakly—the child be very delicate, there is but slight hope of recovery. The danger of the disease is not always to be measured by the state of the throat. Sometimes, when the patient appears to be getting well, a sudden change for the worse rapidly carries him off. Hence the importance of great caution, in such cases, in giving an opinion as to ultimate recovery. I have said enough to prove the terrible nature of the disease, and to show the necessity of calling in, at the earliest period of the symptoms, an experienced and skilful medical man.

209. *Is Diphtheria contagious?*

Decidedly. Therefore, when practicable, the rest of the children ought instantly to be removed to a distance. I say *children,* for it is emphatically a disease of childhood. When adults have it, it is the exception and not the rule : " Thus it will be seen, in the account given of the Boulogne epidemic, that of 366 deaths from this cause, 341 occurred amongst children under ten years of age. In the Lincolnshire epidemic, in the autumn of 1858, all the deaths at Horncastle, 25 in number, occurred amongst children under twelve years of age. " *

210. *What are the causes of Diphtheria?*

Bad and imperfect drainage ; † want of ventilation ; overflowing privies ; low neighbourhoods in the vicinity of rivers ; stagnant waters ; indeed, everything that vitiates the air, and thus depresses the system, more especially if the weather be close and muggy ; poor and improper food ; and last, though not least, contagion. Bear in mind, too, that a delicate child is much more predisposed to the disease than a strong one.

---

* *Diphtheria:* by Ernest Hart. A valuable pamphlet on the subject. Dr Wade of Birmingham has also written an interesting and useful monograph on Diphtheria. I am indebted to the above authors for much valuable information.

† " Now all my carefully conducted inquiries induce me to believe that the disease comes from drain-poison. All the cases into which I could fully inquire, have brought conviction to my mind that there is a direct law of sequence in some peculiar conditions of atmosphere between diphtheria and bad drainage ; and, if this be proved by subsequent investigations, we may be able to prevent a disease which, in too many cases, our known remedies cannot cure."—W. Carr, Esq., Blackheath. *British Medical Journal,* December 7, 1861.

211. What is the treatment of Diphtheria ?

What to do.—Examine well into the ventilation, for as diphtheria is frequently caused by deficient ventilation, the best remedy is thorough ventilation. Look well both to the drains and to the privies, and see that the drains from the water-closets and from the privies do not in any way contaminate the pump-water. If the drains be defective or the privies be full, the disease in your child will be generated, fed, and fostered. Not only so, but the disease will spread in your family and all around you.

Keep the child to his bedroom and to his bed. For the first two or three days, while the fever runs high, put him on a low diet, such as milk, tea, arrow-root, &c.

Apply to his throat every four hours a warm barm and oatmeal poultice. If he be old enough to have the knowledge to use a gargle, the following will be found serviceable :—

> Take of—Permanganate of Potash, pure, four grains ;
> Water, eight ounces :
> To make a Gargle. ᐧ

Or,

> Take of—Powdered Alum, one drachm ;
> Simple Syrup, one ounce ;
> Water, seven ounces :
> To make a Gargle.

The best medicine for the first few days of the attack, is the following mixture :—

> Take of—Chlorate of Potash, two drachms ;
> Boiling Water, seven ounces ;
> Syrup of Red Poppy, one ounce :
> To Make a mixture. A table-spoonful to be taken every four hours.

Or, the chlorate of potash might be given in the form of powder :—

> Take of—Chlorate of Potash, two scruples ;
> Lump Sugar one, drachm :

Mix, and divide into eight powders. One to be put into a dry tea-spoon and then placed on the tongue every three hours. These powders are very useful in diphtheria ; they are very cleansing to the tongue and throat. If they produce much smart ing, as where the mouth is very sore they sometimes do, let the patient, after taking one, drink plentifully of milk; indeed, I have known these powders induce a patient to take nourishment, in the form of milk, which he otherwise would not have done, and thus to have saved him from dying of starvation, which, before taking the powders, there was every probability of his doing. An extensive experience has demonstrated to me the great value of these powders in diphtheria; but they must be put on the tongue dry.

As soon as the skin has lost its preternatural heat, beef-tea and chicken-broth ought to be given. Or if great prostration should supervene, in addition to the beef-tea, port wine, a table-spoonful every four hours, should be administered. If the child be cold, and there be great sinking of the vital powers, brandy and water should be substituted for the port wine. Remember, in ordinary cases, port wine and brandy are not necessary; *but in cases of extreme exhaustion* they are most valuable.

As soon as the great heat of the skin has abated and the debility has set in, one of the following mixtures will be found useful :—

> Take of—Wine of Iron, one ounce and a half;
> Simple Syrup, one ounce ;
> Water, three ounces and a half :
> To make a Mixture. A table-spoonful to be taken every four hours.

Or,

> Take of—Tincture of Perchloride of Iron, one drachm ;
> Simple Syrup, one ounce ;
> Water, three ounces :
> To make a Mixture. A table-spoonful to be taken three times a day.

If the disease should travel downwards, it will cause all the symptoms of croup, then it must be treated as croup; with this only difference, that a blister (*Tela Vesicatoria*) must *not* be applied, or the blistered surface may be attacked by the membrane of diphtheria, which may either cause death or hasten that catastrophe. In every other respect treat the case as croup, by giving an emetic, a tea-spoonful of Ipecacuanha Wine every five minutes, until free vomiting be excited, and then administer smaller doses of Ipecacuanha Wine every two or three hours, as I recommended when conversing with you on the treatment of croup.

What NOT *to do.*—Do not, on any account, apply either leeches or a blister. If the latter be applied, it is almost sure to be covered with the membrane of diph- theria, similar to that inside of the mouth and of the throat, which would be a serious complication. Do not give either calomel or emetic tartar. Do not depress the system by aperients, for diphtheria is an awfully depress- ing complaint of itself ; the patient, in point of fact, is labouring under the depressing effects of poison, for the blood has been poisoned either by the drinking water being contaminated by fæcal matter from either a privy or from a water-closet ; by some horrid drain ; by proximity to a pig-sty ; by an overflowing privy, espe- cially if vegetable matter be rotting at the same time in it ; by bad ventilation, or by contagion. Diphtheria may generally be traced either to the one or to the other of the above causes ; therefore let me urgently entreat you to look well into all these matters, and thus to stay the pes- tilence ! Diphtheria might long remain in a neighbour- hood if active measures be not used to exterminate it.

212. *Have the goodness to describe the symptoms of Measles ?*

Measles commences with symptoms of a common cold ; the patient is at first chilly, then hot and feverish ; he has a running at the nose, sneezing, watering, and redness of the eyes, headache, drowsiness, a hoarse and peculiar ringing cough, which nurses call "measle- cough," and difficulty of breathing. These symptoms usually last three days before the eruption appears ; on the fourth it (the eruption) generally makes its appear- ance, and continues for four days and then disappears, lasting altogether, from the commencement of the symp- toms of cold to the decline of the eruption, seven days. It is important to bear in mind that the eruption con- sists of *crescent-shaped—half-moon-shaped—patches ;* that they usually appear first about the face and the

neck, in which places they are the best marked ; then
on the body and on the arms ; and, lastly, on the legs,
and that they are slightly raised above the surface of the
skin. The face is swollen, more especially the eye-lids
which are sometimes for a few days closed.

Well, then, remember, *the running at the nose, the
sneezing, the peculiar hoarse cough, and the half-moon-
shaped patches*, are the leading features of the disease,
and point out for a certainty that it is measles.

213. *What constitutes the principal danger in Measles ?*

The affection of the chest. The mucous or lining
membrane of the bronchial tubes is always more or less
inflamed, and the lungs themselves are sometimes
affected.

214. *Do you recommend " surfeit water " and saffron
tea to throw out the eruption in Measles ?*

Certainly not. The only way to throw out the eruption, as it is called, is to keep the body comfortably
warm, and to give the beverages ordered by the medical
man, with the chill off. " Surfeit water, " saffron tea,
and remedies of that class, are hot and stimulating.
The only effect they can have, will be to increase the
fever and the inflammation—to add fuel to the fire.

215. *What is the treatment of Measles?*

What to do.—The child ought to be confined both to
his room and to his bed, the room being kept comfortably warm ; therefore, if it be winter time, there should
be a small fire in the grate ; in the summer time, a fire
would be improper. The child must not be exposed to
draughts ; notwithstanding, from time to time, the door
ought to be left a little ajar in order to change the air
of the apartment ; for proper ventilation, let the disease
be what it may, is absolutely necessary.

Let the child, for the first few days, be kept on a low
diet, such as on milk and water, arrow-root, bread and
butter, &c.

If the attack be mild, that is to say, if the breathing
be not much affected (for in measles it always is more
or less affected), and if there be not much wheezing, the

Acidulated Infusion of Roses' Mixture* will be all that is necessary.

But suppose that the breathing is short, and that there is a great wheezing, then instead of giving him the mixture just advised, give him a tea-spoonful of a mixture composed of Ipecacuanha Wine, Syrup, and Water, † every four hours. And if, on the following day, the breathing and the wheezing be not relieved, in addition to the Ipecacuanha Mixture, apply a Tela Vesicatoria, as advised under the head of Inflammation of the Lungs.

When the child is convalescing, batter-puddings, rice, and sago-puddings, in addition to the milk, bread and butter, &c., should be given; and, a few days later, chicken, mutton chops, &c.

The child ought not, even in a mild case of measles, and in favourable weather, to be allowed to leave the house under a fortnight, or it might bring on an attack of bronchitis.

What NOT *to do.*—Do not give either " surfeit water" or wine. Do not apply leeches to the chest. Do not expose the child to the cold air. Do not keep the bed-room very hot, but comfortably warm. Do not let the child leave the house, even under favourable circum-stances, under a fortnight. Do not, while the eruption is out, give aperients. Do not, " to ease the cough," administer either emetic tartar or paregoric—the former drug is awfully depressing; the latter will stop the cough, and will thus prevent the expulsion of the phlegm.

216. *What is the difference between Scarlatina and Scarlet Fever?*

They are indeed one and the same disease, scarlatina being the Latin for scarlet fever. But, in *a popular* sense, when the disease is mild, it is usually called scar-latina. The latter term does not sound so formidable to the ears either of patients or of parents.

---

* See page 178.   † See page 161.

217. *Will you describe the symptoms of Scarlet Fever ?*
The patient is generally chilly, languid, drowsy,
feverish, and poorly for two days before the eruption
appears. At the end of the second day, the characteristic, bright scarlet efflorescence, somewhat similar to
the colour of a boiled lobster, usually first shows itself.
The scarlet appearance is not confined to the skin ; but
the tongue, the throat, and the whites of the eyes put
on the same appearance ; with this only difference, that
on the tongue and on the throat the scarlet is much
darker; and, as Dr Elliotson accurately describes it,—
"the tongue looks as if it had been slightly sprinkled
with Cayenne pepper;" the tongue, at other times,
looks like a strawberry ; when it does, it is called "the
strawberry tongue." The eruption usually declines on
the fifth, and is generally indistinct on the sixth day ;
on the seventh it has completely faded away. There is
usually, after the first few days, great itching on the
surface of the body. The skin, at the end of the week,
begins to peel and to dust off, making it look as though
meal had been sprinkled upon it.

There are three forms of scarlet fever ;—the one where
the throat is little, if at all, affected, and this is a mild
form of the disease ; the second, which is generally,
especially at night, attended with delirium, where the
throat is *much* affected, being often greatly inflamed and
ulcerated ; and the third (which is, except in certain
unhealthy districts, comparatively rare, and which is
VERY dangerous), the malignant form.

218. *Would it be well to give a little cooling, opening
physic as soon as a child begins to sicken for Scarlet Fever ?*
On no account whatever. Aperient medicines are, in
my opinion, highly improper and dangerous both before
and during the period of the eruption. It is my firm
conviction, that the administration of opening medicine,
at such times, is one of the principal causes of scarlet
fever being so frequently fatal. This is, of course, more
applicable to the poor, and to those who are unable to
procure a skilful medical man.

219. *What constitutes the principal danger in Scarlet Fever ?*

The affection of the throat, the administration of opening medicine during the first ten days, and a peculiar disease of the kidneys ending in *anasarca* (dropsy) ; on which account, the medical man ought, when practicable, to be sent for at the onset, that no time may be lost in applying *proper* remedies.

When Scarlet Fever is complicated—as it sometimes is—with diphtheria, the diphtheric membrane is very apt to travel into the wind-pipe, and thus to cause diphtheric croup ; it is almost sure, when such is the case, to end in death. When a child dies from such a complication, the death might truly be said to be owing to the diphtheric croup, and not to the Scarlet Fever ; for if the diphtheric croup had not occurred, the child would, in all probability, have been saved. The deaths from diphtheria are generally from diphtheric croup ; if there be no croup, there is, as a rule, frequent recovery.

220. *How would you distinguish between Scarlet Fever and Measles?*

Measles commences with symptoms of a common cold ; scarlet fever does not. Measles has a *peculiar hoarse* cough ; scarlet fever has not. The eruption of measles is in patches of a half-moon shape, and is slightly raised above the skin ; the eruption of scarlet fever is *not* raised above the skin at all, and is one continued mass. The colour of the eruption is much more vivid in scarlet fever than in measles. The chest is the part principally affected in measles, and the throat in scarlet fever.

There is an excellent method of determining, for a certainty, whether the eruption be that of scarlatina or otherwise. I myself have, in several instances, ascertained the truth of it :—" For several years M. Bouchut has remarked in the eruptions of scarlatina a curious phenomenon, which serves to distinguish this eruption from that of measles, erythema, erysipelas &c., a phenomenon essentially vital, and which is connected with the excessive contractability of the capillaries. The phenomenon in question is a *white line*, which can be produced at pleasure by drawing the back of the nail along the skin where the eruption, is situated. On drawing the nail, or the extremity of a hard body (such as a pen-holder), along the eruption, the skin is observed to grow pale, and to present a white trace, which remains for one or two minutes, or longer, and then disappears. In this way the diagnosis of the disease may be very distinctly written on the skin ; the word ' Scarlatina ' disappears as the eruption regains its uniform tint."—*Edinburgh Medical Journal.*

221. *Is it of so much importance, then, to distinguish between Scarlet Fever and Measles ?*

It is of great importance, as in measles the patient ought to be kept *moderately* warm, and the drinks should be given with the chill off ; while in scarlet fever the patient ought to be kept cool—indeed, for the first few days, *cold ;* and the beverages, such as spring-water, toast and water, &c., should be administered quite cold.

222. *Do you believe in " Hybrid" Scarlet Fever—that is to say, in a cross between Scarlet Fever and Measles ?*

I never in my life saw a case of "hybrid" scarlet fever—nor do I believe in it. Scarlet fever and measles are both blood poisons, each one being perfectly separate and distinct from the other. "Hybrid" scarlet fever is, in my opinion, an utter impossibility. In olden times, when the symptoms of diseases were not so well and carefully distinguished as now, scarlet fever and measles were constantly confounded one with the other, and was frequently said to be "hybrid"—a cross between measles and scarlet fever—to the patient's great detriment and danger, the two diseases being as distinct and separate as their treatment and management ought to be.

223. *What is the treatment of Scarlet Fever ?* *

---

* On the 4th of March 1856, I had the honour to read a *Paper on the Treatment of Scarlet Fever* before the members of

What to do.—Pray pay attention to my rules, and carry out my directions to the letter—I can then promise, *that if the scarlet fever be neither malignant nor complicated with diphtheria,* the plan I am about to advise will, with God's blessing, be usually successful. What is the first thing to be done? Send the child to bed; throw open the windows, be it winter or summer, and have a thorough ventilation; for the bedroom must be kept cool, I may say cold. Do not be afraid of fresh air, for fresh air, for the first few days, is essential to recovery. *Fresh air, and plenty of it, in scarlet fever, is the best doctor* a child can have: let these words be written legibly on your mind.*

If the weather be either intensely cold, or very damp, there is no objection to a small fire in the grate, provided there be, at the same time, air—an abundance of fresh air—admitted into the room.

Take down the curtains of the bed; remove the valances. If it be summer-time, let the child be only covered with a sheet: if it be winter-time, in addition to the sheet, he should have one blanket over him.

Now for the throat.—The best *external* application is a barm and oatmeal poultice. How ought it to be

Queen's College Medico-Chirugical Society, Birmingham,—which *Paper* was afterwards published in the *Association Journal* (March 15, 1856); and in Braithwaite's *Retrospect of Medicine* (January—June, 1856); and in Ranking's *Half-Yearly Abstract of the Medical Sciences* (July—December, 1856); besides in other publications. Moreover, the *Paper* was translated into German, and published in *Canstatt's Jahresbericht,* iv. 456. 1859.

* In the *Times* of Sept. 4, 1863, is the following, copied from the *Bridgewater Mercury:*—
"GROSS SUPERSTITION.—In one of the streets of Taunton, there resides a man and his wife who have the care of a child. This child was attacked with scarlatina, and to all appearance death was inevitable. A jury of matrons was, as it were, empanelled, and to prevent the child 'dying hard,' all the doors in the house, all the drawers, all the boxes, all the cupboards were thrown wide open, the keys taken out, and the body of the child placed under a beam, whereby a sure, certain, and easy passage into eternity could be secured. Watchers held their vigils

made, and how applied? Put half a tea-cupful of barm into a saucepan, put it on the fire to boil; as soon as it boils, take it off the fire, and stir oatmeal into it, until it be of the consistence of a nice soft poultice; then place it on a rag, and apply it to the throat; carefully fasten it on with a bandage, two or three turns of the bandage going round the throat, and two or three over the crown of the head, so as nicely to apply the poultice where it is wanted—that is to say, to cover the tonsils. Tack the bandage : do not pin it. Let the poultice be changed three times a day. The best medicine is the Acidulated Infusion of Roses, sweetened with syrup :—

> Take of—Diluted Sulphuric Acid, half a drachm ;
> Simple Syrup, one ounce and a half ;
> Acid Infusion of Roses, four ounces and a half:
> To make a Mixture. A table-spoonful to be taken every four hours.

It is grateful and refreshing, it is pleasant to take, it abates fever and thirst, it cleanses the throat and tongue of mucus, and is peculiarly efficacious in scarlet fever ; as soon as the fever is abated it gives an appetite. My belief is that the sulphuric acid in the mixture is a specific in scarlet fever, as much as quinine is in ague, and sulphur in itch. I have reason to say so, for, in numerous cases I have seen its immense value.

---

throughout the weary night, and in the morning the child, to the surprise of all, did not die, and is now gradually recovering."

These old women—this jury of matrons—stumbled on the right remedy, "all the doors in the house . . . . were thrown wide open," and thus they thoroughly ventilated the apartment. What was the consequence? The child who, just before the opening of the doors, had all the appearances "that death was inevitable," as soon as fresh air was let in showed symptoms of recovery, "and in the morning the child, to the surprise of all, did not die, and is now gradually recovering." There is nothing wonderful—there is nothing surprising to my mind— in all this. Ventilation—thorough ventilation—is the grand remedy for scarlatina ! Oh, that there were in scarlet fever cases a good many such old women's—such a "jury of matrons'" —remedies ! We should not then be horrified, as we now are, at the fearful records of death, which the Returns of the Registrar-General disclose !

Now, with regard to food.—If the child be at the breast, keep him entirely to it. If he be weaned, and under two years old, give him milk and water, and cold water to drink. If he be older, give him toast and water, and plain water from the pump, as much as he chooses ; let it be quite cold—the colder the better. Weak black tea, or thin gruel, may be given, but not caring, unless he be an infant at the breast, if he take nothing but *cold* water. If the child be two years old and upwards, roasted apples with sugar, and grapes, will be very refreshing, and will tend to cleanse both the mouth and the throat. Avoid broths and stimulants.

When the appetite returns, you may consider the patient to be safe. The diet ought now to be gradually improved. Bread and butter, milk and water, and arrow-root made with equal parts of new milk and water, should for the first two or three days be given. Then a light batter or rice pudding may be added, and in a few days, either a little chicken or a mutton chop.

The essential remedies, then, in scarlet fever, are, for the first few days—(1) plenty of fresh air and ventilation, (2) plenty of cold water to drink, (3) barm poultices to the throat, and (4) the Acidulated Infusion of Roses Mixture as a medicine.

Now, then, comes very important advice. After the first few days, probably five or six, sometimes as early as the fourth day—*watch carefully and warily, and note the time, the skin will suddenly become cool,* the child will say that he feels chilly ; then is the time you must now change your tactics—*instantly close the windows and put extra clothing,* a blanket or two, on his bed. A flannel night-gown should, until the dead skin have peeled off, be now worn next to the skin, when the flannel night-gown should be discontinued. The patient ought ever after to wear, in the day time, a flannel waist-coat.* His drinks must now be given with the chill off ;

---

* On the importance—the vital importance—of the wearing of flannel next to the skin, see " Flannel Waistcoats "

he ought to have a warm cup of tea, and gradually his
diet should, as I have previously advised, be improved.

There is one important caution I wish to impress upon
you,—*do not give opening medicine during the time the
eruption is out.* In all probability the bowels will be
opened : if so, all well and good ; but do not, on any
account, for the first ten days, use artificial means to
open them. It is my firm conviction that the administration of purgatives in scarlet fever is a fruitful source
of dropsy, of disease, and death. When we take into
consideration the sympathy there is between the skin
and the mucous membrane, I think that we should
pause before giving irritating medicines, such as purgatives. The irritation of aperients on the mucous
membrane may cause the poison of the skin disease (for
scarlet fever is a blood-poison) to be driven internally to
the kidneys, to the throat, to the pericardium (bag of
the heart), or to the brain. You may say, Do you not
purge if the bowels be not open for a week? I say
emphatically, No !

I consider my great success in the treatment of scarlet
fever to be partly owing to my avoidance of aperients
during the first ten days of the child's illness.

If the bowels, after the ten days, be not properly
opened, a dose or two of syrup of senna should be given :
that is to say, one or two tea-spoonfuls should be administered early in the morning, and should, if the first dose
does not operate, be repeated in four hours.

In a subsequent Conversation, I shall strongly urge
you not to allow your child, when convalescent, to leave
the house under at least a month from the commencement of the illness ; I, therefore, beg to refer you to that
Conversation, and hope that you will give it your best
and earnest consideration ! During the last twenty
years I have never had dropsy from scarlet fever, and I
attribute it entirely to the plan I have just recommended, and in not allowing my patients to leave the
house under the month—until, in fact, the skin that
had peeled off has been renewed,

Let me now sum up the plan I adopt, and which I beg leave to designate as—Pye Chavasse's Fresh Air Treatment of Scarlet Fever :—

1. Thorough ventilation, a cool room, and scant clothes on the bed, for the first five or six days.

2. A change of temperature of the skin to be carefully regarded. As soon as the skin is cool, closing the windows, and putting additional clothing on the bed.

3. The Acidulated Infusion of Roses with Syrup is *the* medicine for scarlet fever.

4. Purgatives to be religiously avoided for the first ten days at least, and even afterwards, unless there be absolute necessity.

5. Leeches, blisters, emetics, cold and tepid spongings, and painting the tonsils with caustic, inadmissible in scarlet fever.

6. A strict antiphlogistic (low) diet for the first few days, during which time cold water to be given *ad libitum.*

7. The patient *not* to leave the house in the summer under the month ; in the winter, under six weeks.

What NOT *to do.*—Do not, then, apply either leeches or blisters to the throat ; do not paint the tonsils with caustic ; do not give aperients ; do not, on any account, give either calomel or emetic tartar ; do not, for the first few days of the illness, be afraid of *cold air* to the skin, and of cold water as a beverage ; do not, emphatically let me say, *do not* let the child leave the house for at least a month from the commencement of the illness.

My firm conviction is, that purgatives, emetics, and blisters, by depressing the patient, sometimes cause ordinary scarlet fever to degenerate into malignant scarlet fever.

I am aware that some of our first authorities advocate a different plan to mine. They recommend purgatives, which I may say, in scarlet fever, are my dread and abhorrence. They advise cold and tepid spongings—a plan which I think dangerous, as it will probably drive the disease internally. Blisters, too, have been prescribed ; these I consider weakening, injurious, and barbarous, and likely still more to inflame the already inflamed skin. They recommend leeches to the throat, which I am convinced, by depressing the patient, will lessen the chance of his battling against the disease, and will increase the ulceration . of the tonsils. Again, the patient has not too much blood ; the blood is only poisoned. I look upon scarlet fever as a specific poison of the blood, and one which will be eliminated from the system, *not* by bleeding, *not* by purgatives, *not* by emetics, but by a constant supply of fresh and cool air, by the acid treatment, by cold water as a beverage, and for the first few days by a strict antiphlogistic (low) diet. Sydenham says that scarlet fever is oftentimes " fatal through the officiousness of the doctor." I conscientiously believe that a truer remark was never made ; and that, under a different system to the usual one adopted, scarlet fever would not be so much dreaded.*

Dr Budd, of Bristol, recommends, in the *British Medical Journal*, that the body, including the scalp, of a scarlet fever patient, should, after about the fourth day, be anointed, every night and morning, with camphorated oil ; this anointing to be continued until the patient is able to take a warm bath and use disinfectant soap : this application will not only be very agreeable to the patient's feelings, as there is usually great irritation and itching of the skin, but it will, likewise, be an important

---

* If any of my medical brethren should do me the honour to read these pages, let me entreat them to try my plan of treating scarlet fever, as my success has been great. I have given full and minute particulars, in order that they and mothers (if mothers cannot obtain medical advice) may give my plan a fair and impartial trial. My only stipulations are that they must *begin* with my treatment, and *not mix* any other with it, and carry out my plan to the very letter. I then, with God's blessing, provided the cases be neither malignant nor complicated with diphtheria, shall not fear the result. If any of my *confrères* have tried my plan of treatment of scarlet fever—and I have reason to know that many have—I should feel grateful to them if they would favour me with their opinion as to its efficacy. Address—" Pye Chavasse, 214 Hagley Road, Birmingham."

means of preventing the dead skin, which is highly infectious, and which comes off partly in flakes and partly floats about the air as dust, from infecting other persons. The plan is an excellent one, and cannot be too strongly recommended.

If the case be a combination of scarlet fever and of diphtheria, as it unfortunately now frequently is, let it be treated as a case of diphtheria.

224. *I have heard of a case of Scarlet Fever, where the child, before the eruption showed itself, was suddenly struck prostrate, cold, and almost pulseless : what, in such a case, are the symptoms, and what immediate treatment do you advise ?*

There is an *exceptional* case of scarlet fever, which now and then occurs, and which requires *exceptional* and prompt treatment, or death will quickly ensue. We will suppose a case : one of the number, where nearly all the other children of a family are labouring under scarlet fever, is quite well, when suddenly—in a few hours, or even, in some cases, in an hour—utter prostration sets in, he is very cold, and is almost pulseless, and is nearly insensible—comatose.

Having sent instantly for a judicious medical man, apply, until he arrives, hot bottles, hot bricks, hot bags of salt to the patient's feet and legs and back, wrap him in hot blankets, close the window, and give him hot brandy and water—a tablespoonful of brandy to half a tumblerful of hot water—give it him by teaspoonfuls, continuously—to keep him alive ; when he is warm and restored to consciousness, the eruption will probably show itself, and he will become hot and feverish ; then your tactics must, at once, be changed, and my Fresh Air Treatment, and the rest of the plan I have before advised must in all its integrity; be carried out.

We sometimes hear of a child, before the eruption comes out and within twenty-four hours of the attack, dying of scarlet fever. When such be the case it is probably owing to low vitality of the system—to utter prostration—he is struck down, as though for death, and if the plan be not adopted of, for a few hours, keeping him alive by heat, and by stimulants, until, indeed, the eruption comes out, he will never rally again, but will die from scarlet fever poisoning and from utter exhaustion. These cases are comparatively rare, but they do, from time to time, occur, and, when they do, they demand exceptional and prompt and energetic means to save them from ending in almost immediate and certain death. " To be forewarned is to be forearmed."*

225. *How soon ought a child to be allowed to leave the house after an attack of Scarlet Fever?*

He must *not* be allowed to go out for at least a month from the commencement of the attack, in the summer, and six weeks in the winter ; and not even then without the express permission of a medical man. It might be said that this is an unreasonable recommendation : but when it is considered that the whole of the skin generally desquamates, or peels off, and consequently leaves the surface of the body exposed to cold, which cold flies to the kidneys, producing a peculiar and serious disease in them, ending in dropsy, this warning will not be deemed unreasonable.

Scarlet fever dropsy, which is really a *formidable disease, generally arises from the carelessness, the ignorance, and the thoughtlessness of parents in allowing a child to leave the house before the new skin be properly formed and hardened.* Prevention is always better than cure.

Thus far with regard to the danger to the child himself. Now, if you please, let me show you the risk of contagion that you inflict upon families, in allowing your child to mix with others before a month at least has elapsed. Bear in mind, a case is quite as contagious, if not more so, while the skin is peeling off, as it was

---

* I have been reminded of this *exceptional* case of scarlet fever by a most intelligent and valued patient of mine, who had a child afflicted as above described, and whose child was saved from almost certain death, bv a somewhat similar plan of treatment as advised in the text.

before. Thus, in ten days or a fortnight, there is as much risk of contagion as at the *beginning* of the disease, and when the fever is at its height. At the conclusion of the month, the old skin has generally all peeled off, and the new skin has taken its place ; consequently there will then be less fear of contagion to others. But the contagion of scarlet fever is so subtle and so uncertain in its duration, that it is impossible to fix the exact time when it ceases.

Let me most earnestly implore you to ponder well on the above important facts. If these remarks should be the means of saving only one child from death, or from broken health, my labour will not have been in vain.

226. *What means do you advise to purify a house, clothes, and furniture, from the contagion of Scarlet Fever ?*

Let every room in the house, together with its contents, and clothing and dresses that cannot be washed, be well fumigated with sulphur—taking care the while to close both windows and door ; let every room be *lime-washed* and then be white-washed ; if the contagion have been virulent, let every bedroom be freshly papered (the walls having been previously stripped of the old paper and then lime-washed) ; let the bed, the bolsters, the pillows, and the mattresses be cleansed and purified ; let the blankets and coverlids be thoroughly washed, and then let them be exposed to the open air—if taken into a field so much the better ; let the rooms be well scoured ; let the windows, top and bottom, be thrown wide open ; let the drains be carefully examined ; let the pump water be scrutinised, to see that it be not contaminated by fæcal matter, either from the water-closet, from the privy, from the pig-stye, or from the stable ; let privies be emptied of their contents—*remember this is most important advice*—then put, into the empty places, either lime and powdered charcoal or carbolic acid, for it is a well-ascertained fact that it is frequently impossible to rid a house of the infection of scarlet fever without adopting such a course. " In St George's, Southwark, the medical officer reports that scarlatina ' has raged fatally, almost exclusively where privy or drain smells are to be perceived in the houses.' "* Let the children, who have not had, or who do not appear to be sickening for scarlet fever, be sent away from home—if to a farm house so much the better. Indeed, leave no stone unturned, no means untried, to exterminate the disease from the house and from the neighbourhood. Remember the young are more prone to catch contagious diseases than adults ; for

> "in the morn and liquid dew of youth ¶
> Contagious blastments are most imminent."—*Shakspeare.*

227. *Have you any further observations to offer on the precautions to be taken against the spread of Scarlet Fever ?*

Great care should be taken to separate the healthy from the infected. The nurses selected for attending scarlet fever patients should be those who have previously had scarlet fever themselves. Dirty linen should be removed at once, and be put into boiling water. Very little furniture should be in the room of a scarlet fever patient—the less the better—it only obstructs the circulation of the air, and harbours the scarlet fever poison. The most scrupulous attention to cleanliness should, in these cases, be observed. A patient who has recovered from scarlet fever, and before he mixes with healthy people, should, for three or four consecutive mornings, have a warm bath, and well wash himself, while in the bath, with soap ; he will, by adopting this plan, get rid of the dead skin, and thus remove the infected particles of the disease. If scarlet fever should appear in a school, the school must for a time be broken up, in order that the disease might be stamped out. There must be no half measures where such a fearful disease is in question. A house containing scarlet fever patients should, by parents, be avoided as the plague ; it is a folly at any time to put one's head into the lion's

---

* * *Quarterly Report of the Board of Health* upon Sickness in the Metropolis.

mouth! Chloralum and carbolic acid, and chloride of lime, and Condy's fluid, are each and all good disinfectants; but not one is to be compared to perfect cleanliness and to an abundance of fresh and pure air—the last of which may truly *par excellence* be called God's disinfectant! Either a table-spoonful of chloralum, or two tea-spoonfuls of carbolic acid, or two tea-spoonfuls of Condy's fluid, or a tea-spoonful of chloride of lime in a pint of water, are useful to sprinkle the soiled handkerchiefs as soon as they be done with, and before they be washed, to put in the *pot-de-chambre*, and to keep in saucers about the room; but, remember, as I have said before, and cannot repeat too often, there is no preventative like the air of heaven, which should be allowed to permeate and circulate freely through the apartment and through the house : air, air, air is the best disinfectant, curative, and preventative of scarlet fever in the world !

I could only wish that my *Treatment of Scarlet Fever* were, in all its integrity, more generally adopted ; if it were, I am quite sure that thousands of children would annually be saved from broken health and from death. Time still further convinces me that my treatment is based on truth, as I have every year additional proofs of its value and of its success ; but error and prejudice are unfortunately ever at work, striving all they can to defeat truth and common sense. One of my principal remedies in the treatment of scarlet fever is an abundance of fresh air; but many people prefer their own miserable complicated inventions to God's grand and yet simple remedies—they pretend that they know better than the Mighty Framer of the universe !

228. *Will you describe the symptoms of Chicken-pox ?*

It is occasionally, but not always, ushered in with a slight shivering fit; the eruption shows itself in about twenty-four hours from the child first appearing poorly. It is a vesicular* disease. The eruption comes out in

---

* *Vesicles.* Small elevations of the cuticle, covering a fluid

the form of small pimples, and principally attacks the scalp, the neck, the back, the chest, and the shoulders, but rarely the face ; while in small-pox the face is generally the part most affected. The next day these pimples fill with water, and thus become vesicles ; on the third day they are at maturity. The vesicles are quite separate and distinct from each other. There is a slight redness around each of them. Fresh ones, whilst the others are dying away, make their appearance. Chicken-pox is usually attended with a slight itching of the skin ; when the vesicles are scratched the fluid escapes, and leaves hard pearl-like substances, which, in a few days, disappear. Chicken-pox never leaves pit marks behind. It is a child's complaint ; adults scarcely, if ever, have it.

229. *Is there any danger in Chicken-pox ; and what treatment do you advise ?*

It is not at all a dangerous, but, on the contrary, a trivial complaint. It lasts only a few days, and requires but little medicine. The patient ought, for three or four days, to keep the house, and should abstain from animal food. On the sixth day, but not until then, a dose or two of a mild aperient is all that will be required.

230. *Is Chicken-pox infectious ?*

There is a diversity of opinion on this head, but one thing is certain—it cannot be communicated by inoculation.

231. *What are the symptoms of Modified Small-pox ?*

The Modified Small-pox—that is to say, small-pox that has been robbed of its virulence by the patient having been either already vaccinated, or by his having had a previous attack of small-pox—is ushered in with severe symptoms, with symptoms almost as severe as though the patient had not been already somewhat protected either by vaccination or by the previous attack

---

which is generally clear and colourless at first, but becomes afterwards whitish and opaque, or pearly.— *Watson.*

of small-pox—that is to say, he has a shivering fit, great depression of spirits and debility, *malaise,* sickness, headache, and occasionally delirium. After the above symptoms have lasted about three days, the eruption shows itself. The immense value of the previous vaccination, or the previous attack of small-pox, now comes into play. In a case of *unprotected* small-pox, the appearance of the eruption *aggravates* all the above symptoms, and the danger begins; while in the *modified* small-pox, the moment the eruption shows itself, the patient feels better, and, as a rule, rapidly recovers. The eruption of *modified* small-pox varies materially from the eruption of the *unprotected* small-pox. The former eruption assumes a varied character, and is composed, first, of vesicles (containing water); and, secondly, of pustules (containing matter), each of which pustules has a depression in the centre; and, thirdly, of several red pimples without either water or matter in them, and which sometimes assume a livid appearance. These "breakings-out" generally show themselves more upon the wrist, and sometimes up one or both of the nostrils. While in the latter disease—the *unprotected* small-pox—the "breaking-out" is composed entirely of pustules containing matter, and which pustules are more on the face than on any other part of the body. There is generally a peculiar smell in both diseases—an odour once smelt never to be forgotten.

Now, there is one most important remark I have to make,—*the modified small-pox is contagious.* This ought to be borne in mind, as a person labouring under the disease must, if there be children in the house, either be sent away himself, or else the children ought to be banished both the house and the neighbourhood. Another important piece of advice is,—let *all* in the house—children and adults, one and all—be vaccinated, even if any or all have been previously vaccinated.

Treatment.—Let the patient keep his room, and if he be very ill, his bed. Let the chamber be well ventilated. If it be winter time, a small fire in the grate will encourage ventilation. If it be summer, a fire is out of the question; indeed, in such a case, the window-sash ought to be opened, as thorough ventilation is an important requisite of cure, both in small-pox and in *modified* small-pox. While the eruption is out, do not on any account give aperient medicine. In ten days from the commencement of the illness a mild aperient may be given. The best medicine in these cases is, the sweetened Acidulated Infusion of Roses,* which ought to be given from the commencement of the disease, and should be continued until the fever be abated. For the first few days, as long as the fever lasts, the patient, ought not to be allowed either meat or broth, but should. be kept on a low diet, such as on gruel, arrow-root, milk-puddings, &c. As soon as the fever is abated he ought gradually to resume his usual diet. When he is convalescent, it is well, where practicable, that he should have change of air for a month.

232. *How would you distinguish between Modified Small-pox and Chicken-pox?*

Modified small-pox may readily be distinguished from chicken-pox, by the former disease being, notwithstanding its modification, much more severe and the fever much more intense *before* the eruption shows itself than chicken-pox; indeed, in chicken-pox there is little or no fever either before or after the eruption; by the former disease—the modified small-pox—consisting *partly* of pustules (containing matter), each pustule having a depression in the centre, and the favourite localities of the pustules being the wrists and the inside of the nostrils; while, in the chicken-pox, the eruption consists of vesicles (containing water), and *not* pustules (containing matter), and the vesicles having neither a depression in the centre, nor having any particular partiality to attack either the wrists or the inside of the nose. In modified small-pox each pustule is, as in unprotected small-pox, inflamed at the base; while in chicken-pox

---

* See page 178.

there is only very slight redness around each vesicle. The vesicles in chicken-pox are small—much smaller than the pustules in modified small-pox.

233. *Is Hooping-cough an inflammatory disease?*

Hooping-cough in itself is not inflammatory, it is purely spasmodic; but it is generally accompanied with more or less of bronchitis—inflammation of the mucous membrane of the bronchial tubes—on which account it is necessary, *in all cases* of hooping-cough, to consult a medical man, that he may watch the progress of the disease and nip inflammation in the bud.

234. *Will you have the goodness to give the symptoms, and a brief history of, Hooping-cough?*

Hooping-cough is emphatically a disease of the young; it is rare for adults to have it; if they do, they usually suffer more severely than children. A child seldom has it but once in his life. It is highly contagious, and therefore frequently runs through a whole family of children, giving much annoyance, anxiety, and trouble to the mother and the nurses; hence hooping-cough is much dreaded by them. It is amenable to treatment. Spring and summer are the best seasons of the year for the disease to occur. This complaint usually lasts from six to twelve weeks—sometimes for a much longer period, more especially if proper means are not employed to relieve it.

Hooping-cough commences as a common cold and cough. The cough, for ten days or a fortnight, increases in intensity; at about which time it puts on the characteristic "hoop." The attack of cough comes on in paroxysms. In a paroxysm, the child coughs so long and so violently, and *expires* so much air from the lungs without *inspiring* any, that at times he appears nearly suffocated and exhausted; the veins of his neck swell; his face is nearly purple; his eyes, with the tremendous exertion, almost seem to start from their sockets; at length there is a sudden *inspiration* of air through the contracted chink of the upper part of the wind-pipe—the glottis—causing the peculiar "hoop;" and after a little

more coughing, he brings up some glairy mucus from
the chest; and sometimes, by vomiting, food from the
stomach; he is at once relieved, until the next paroxysm
occur, when the same process is repeated, the child during
the intervals, in a favourable case, appearing quite well,
and after the cough is over, instantly returning either to
his play or to his food. Generally, after a paroxysm he
is hungry, unless, indeed, there be severe inflammation
either of the chest or of the lungs. Sickness, as I before
remarked, frequently accompanies hooping-cough; when
it does, it might be looked upon as a good sign. The
child usually knows when an attack is coming on; he
dreads it, and therefore tries to prevent it; he sometimes
partially succeeds; but, if he does, it only makes the
attack, when it does come, more severe. All causes of
irritation and excitement ought, as much as possible, to
be avoided, as passion is apt to bring on a severe
paroxysm.

A new-born babe—an infant of one or two months old
—commonly escapes the infection; but if, at that tender
age, he unfortunately catch hooping-cough, it is likely to
fare harder with him than if he were older—the younger
the child, the greater the risk. But still, in such a case,
do not despair, as I have known numerous instances of
new-born infants, with judicious care, recover perfectly
from the attack, and thrive after it as though nothing of
the kind had ever happened.

A new-born babe, labouring under hooping-cough, is
liable to convulsions, which is in this disease one, indeed
the great, source of danger. A child, too, who is teething, and labouring under the disease, is also liable to
convulsions. When the patient is convalescing, care
ought to be taken that he does not catch cold, or the
"hoop" might return. Hooping-cough may either
precede, attend, or follow an attack of measles.

235. *What is the treatment of Hooping-cough?*

We will divide the hooping-cough into three stages,
and treat each stage separately,

What to do.—In the first stage, the commencement of

hooping-cough: For the first ten days give the Ipecacuanha Wine Mixture,* a tea-spoonful three times a day. of the child be not weaned, keep him entirely to the breast; if he be weaned, to a milk and farinaceous diet. Confine him for the first ten days to the house, more especially if the hooping-cough be attended, as it usually is, with more or less bronchitis. But take care that the rooms be well ventilated; for good air is essential to the cure.

If the bronchitis attending the hooping-cough be severe, confine him to his bed, and treat him as though it were simply a case of bronchitis.†

In the second stage, discontinue the Ipecacuanha Mixture, and give Dr Gibb's remedy—namely, Nitric Acid—which I have found to be an efficacious and valuable one in hooping-cough :—

> Take of—Diluted Nitric Acid, two drachms;
> Compound Tincture of Cardamons, half a drachm;
> Simple Syrup, three ounces;
> Water, two ounces and a half:
>
> Make a Mixture. One or two tea-spoonfuls, or a table-spoonful, according to the age of the child—one tea-spoonful for an infant of six months, and two tea-spoonfuls for a child of twelve months, and one table-spoonful for a child of two years, every four hours, first shaking the bottle.

Let the spine and the chest be well rubbed every night and morning either with Roche's Embrocation, or with the following stimulating liniment (first shaking the bottle) :—

> Take of—Oil of Cloves, one drachm;
> Oil of Amber, two drachms;
> Camphorated Oil, nine drachms:
> Make a Liniment.

Let him wear a broad band of new flannel, which should extend round from his chest to his back, and which ought to be changed every night and morning, in

---

* For the prescription of the Ipecacuanha Wine Mixture, see page 161.

† For the treatment of bronchitis, see page 164.

A

order that it may be dried before putting on again. To
keep it in its place it should be fastened by means of
tapes and with shoulder-straps.

The diet ought now to be improved—he should
gradually return to his usual food; and, weather permitting, should almost live in the open air—fresh air
being, in such a case, one of the finest medicines.

In the third stage, that is to say, when the complaint
has lasted a month, if by that time the child is not well,
there is nothing like change of air to a high, dry, healthy,
country place. Continue the Nitric Acid Mixture, and
either the Embrocation or the Liniment to the back and
the chest, and let him continue to almost live in the
open air, and be sure that he does not discontinue wearing the flannel until he be quite cured, and then let it
be left off by degrees.

If the hooping-cough have caused debility, give him
Cod-liver Oil—a tea-spoonful twice or three times a day,
giving it him on a full stomach, after his meals. But,
remember, after the first three or four weeks, change of
air, and plenty of it, is for hooping-cough the grand
remedy.

What NOT *to do.*—Do not apply leeches to the chest,
for I would rather put blood into a child labouring
under hooping-cough than take it out of him—hooping-
cough is quite weakening enough to the system of itself
without robbing him of his life's blood; do not, on any
account whatever, administer either emetic tartar or
antimonial wine; do not give either paregoric or syrup
of white poppies; do not drug him either with calomel
or with grey-powder; do not dose him with quack
medicine; do not give him stimulants, but rather give
him plenty of nourishment, such as milk and
farinaceous food, but *no* stimulants; do not be afraid,
after the first week or two, of his having fresh air, and
plenty of it—for fresh, pure air is the grand remedy,
after all that can be said and done, in hooping-cough.
Although occasionally we find that, if the child be
labouring under hooping-cough, and is breathing a pure

country air, and is not getting well so rapidly as we could wish, change of air to a smoky gas-laden town will sometimes quickly effect a cure ; indeed, some persons go so far as to say that the *best* remedy for an *obstinate* case of hooping-cough is, for the child to live, the great part of every day, in gas-works !

236. *What is to be done during · a paroxysm of Hooping-cough ?*

If the child be old enough, let him stand up ; but if he be either too young or too feeble, raise his head, and bend his body a little forward ; then support his back with one hand, and the forehead with the other. Let the mucus, the moment it be within reach, be wiped with a soft handkerchief out of his mouth.

237. *In an obstinate case of Hooping-cough, what is the best remedy ?*

Change of air, provided there be no active inflammation, to any healthy spot. A farm-house, in a high, dry, and salubrious neighbourhood, is as good a place as can be chosen. If, in a short time, he be not quite well, take him to the sea-side : the sea breezes will often, as if by magic, drive away the disease.

238. *Suppose my child should have a shivering fit, is it to be looked upon as an important symptom ?*

Certainly. Nearly all *serious* illnesses commence with a shivering fit : severe colds, influenza, inflammations of different organs, scarlet fever, measles, small-pox, and very many other diseases, begin in this way. If, therefore, your child should ever have a shivering fit, *instantly* send for · a medical man, as delay might be dangerous. A few hours of judicious treatment, at the commencement of an illness, is frequently of more avail than days and weeks, nay months, of treatment, when disease has gained a firm footing. A *serious* disease often steals on insidiously, and we have perhaps only the shivering fit, which might be but a *slight* one, to tell us of its approach.

A *trifling* ailment, too, by neglecting the premonitory symptom, which, at first, might only be indicated by a *slight* shivering fit, will sometimes become a *mortal* disorder :—

> " The little rift within the lute,
> That by-and-by will make the music mute,
> And ever widening slowly silence all."*

239. *In case of a shivering fit, perhaps you will tell me what to do ?*

Instantly have the bed warmed, and put the child to bed. Apply either a hot bottle or a hot brick, wrapped in flannel, to the soles of his feet. Put an extra blanket on his bed, and give him a cup of hot tea. As soon as the shivering fit is over, and he has become hot, gradually lessen the *extra* quantity of clothes on his bed, and take away the hot bottle or the hot brick from his feet.

What NOT *to do.*—Do not give either brandy or wine, as inflammation of some organ might be about taking place. Do not administer opening medicine, as there might be some " breaking out " coming out on the skin, and an aperient might check it.

240. *My child, apparently otherwise healthy, screams out in the night violently in his sleep, and nothing for a time will pacify him : what is likely to be the cause, and what is the treatment ?*

The causes of these violent screamings in the night are various. At one time, they proceed from teething ; at another, from worms ; sometimes, from night-mare ;

---

* The above extract from Tennyson is, in my humble opinion, one of the most beautiful pieces of poetry in the English language. It is a perfect gem, and a volume in itself, so truthful, so exquisite, so full of the most valuable reflections ; for instance—(1.) " The little rift within the lute,"—the little tubercle within the lung " that by-and-by will make the music mute, and ever widening slowly silence all," and the patient eventually dies of consumption. (2.) The little rent—the little rift of a very minute vessel in the brain, produces an attack of apoplexy, and the patient dies. (3.) Each and all of us, in one form or another, sooner or later, will have " the little rift within the lute." But why give more illustrations ?—a little reflection will bring numerous examples to my fair reader's memory,

occasionally, from either disordered stomach or bowels. Each of the above causes will, of course, require a different plan of procedure; it will, therefore, be necessary to consult a medical man on the subject, who will soon, with appropriate treatment, be able to relieve him.

241. *Have the goodness to describe the complaint of children called Mumps.*

The mumps, inflammation of the "parotid" gland, is commonly ushered in with a slight feverish attack. After a short time, a swelling, of stony hardness, is noticed before and under the ear, which swelling extends along the neck towards the chin. This lump is exceedingly painful, and continues painful and swollen for four or five days. At the end of which time it gradually disappears, leaving not a trace behind. The swelling of mumps never gathers. It may affect one or both sides of the face. It seldom occurs but once in a lifetime. It is contagious, and has been known to run through a whole family or school; but it is not dangerous, unless, which is rarely the case, it leaves the "parotid" gland, and migrates either to the head, to the breast, or to the testicle.

242. *What is the treatment of Mumps?*

Foment the swelling, four or five times a day, with a flannel wrung out of hot camomile and poppy-head decoction;* and apply, every night, a barm and oatmeal poultice to the swollen gland or glands. Debar, for a few days, the little patient from taking meat and broth, and let him live on bread and milk, light puddings, and arrow-root. Keep him in a well-ventilated room, and shut him out from the company of his brothers, his sisters, and young companions. Give him a little mild, aperient medicine. Of course, if there be the slightest symptom of migration to any other part or parts, instantly call in a medical man.

---

*Four poppy-heads and four ounces of camomile blows to be boiled in four pints of water for half an hour, and then strained to make the decoction.

243. *What is the treatment of a Boil?*

One of the best applications is a Burgundy-pitch plaster spread on a soft piece of wash leather. Let a chemist spread a plaster, about the size of the hand; and, from this piece, cut small plasters, the size of a shilling or a florin (according to the dimensions of the boil), which snip around and apply to the part. Put a fresh one on daily. This plaster will soon cause the boil to break; when it does break, squeeze out the contents—the core and the matter—and then apply one of the plasters as before, which, until the boil be well, renew every day.

The old-fashioned remedy for a boil—namely, common yellow soap and brown-sugar, is a capital one for the purpose. It is made with equal parts of brown sugar and of shredded yellow soap, and mixed by means of a table-knife on a plate, with a few drops of water, until it be all well blended together, and of the consistence of thick paste; it should then be spread either on a piece of wash-leather, or on thick linen, and applied to the boil, and kept in its place by means either of a bandage or of a folded handkerchief; and should be removed once or twice a day. This is an excellent application for a boil—soothing, comforting, and drawing—and will soon effect a cure. A paste of honey and flour, spread on linen rag, is another popular and good application for a boil.

If the boils should arise from the child being in a delicate state of health, give him cod-liver oil, meat once a day, and an abundance of milk and farinaceous food. Let him have plenty of fresh air, exercise, and play.

If the boils should arise from gross and improper feeding, then keep him for a time from meat, and let him live principally on a milk and farinaceous diet.

If the child be fat and gross, cod-liver oil would be improper; a mild aperient, such as rhubarb and magnesia, would then be the best medicine.

244. *What are the symptoms of Ear-ache?*

A young child screaming shrilly, violently, and continuously, is oftentimes owing to ear-ache; carefully, therefore, examine each ear, and ascertain if there be any discharge; if there be, the mystery is explained.

Screaming from ear-ache may be distinguished from the screaming from bowel-ache by the former (ear-ache) being more continuous—indeed, being one continued scream, and from the child putting his hand to his head; while, in the latter (bowel-ache), the pain is more of a coming and of a going character, and he draws up his legs to his bowels. Again, in the former (ear-ache), the secretions from the bowels are natural; while, in the latter (bowel-ache), the secretions from the bowels are usually depraved, and probably offensive. But a careful examination of the ear will generally at once decide the nature of the case.

2.5. *Wnat is the best remedy for Ear-ache?*

Apply to the ear a small flannel bag, filled with hot salt—as hot as can be comfortably borne, or foment the ear with a flannel wrung out of hot camomile and poppy head decoction. A roasted onion, inclosed in muslin applied to the ear, is an old-fashioned and favourite remedy, and may, if the bag of hot salt, or if the hot fomentation do not relieve, be tried. Put into the ear, but not very far, a small piece of cotton wool, moistened with warm olive oil. Taking care that the wool is always removed before a fresh piece be substituted, as if it be allowed to remain in any length of time, it may produce a discharge from the ear. Avoid all *cold* applications. If the ear-ache be severe, keep the little fellow at home, in a room of equal temperature, but well-ventilated, and give him, for a day or two, no meat.

If a discharge from the ear should either accompany or follow the ear-ache, *more especially if the discharge be offensive,* instantly call in a medical man, or deafness for life may be the result.

A knitted or crotcheted hat, with woollen rosettes over the ears, is, in the winter time, an excellent hat for a child subject to ear-ache. The hat may be procured at any baby-linen warehouse.

246. *What are the causes and the treatment of discharges from the Ear ?*

Cold, measles, scarlet fever, healing up of " breakings-out " behind the ear ; pellets of cotton wool, which had been put in the ear, and had been forgotten to be removed, are the usual causes of discharges from the ear. It generally commences with ear-ache.

The *treatment* consists in keeping the parts clean, by syringing the ear every morning with warm water, by attention to food—keeping the child principally upon a milk and a farinaceous diet, and by change of air—more especially to the coast. If change of air be not practicable, great attention should be paid to ventilation. As I have before advised, in all cases of discharge from the ear call in a medical man, as a little judicious medicine is advisable—indeed, essential ; and it may be necessary to syringe the ear with lotions, instead of with warm water ; and, of course, it is only a doctor who has actually seen the patient who can decide these matters, and what is best to be done in each case.

247. *What is the treatment of a " stye " on the eye-lid ?*

Bathe the eye frequently with warm milk and water, and apply, every night at bedtime, a warm white-bread poultice.

No medicine is required ; but, if the child be gross, keep him for a few days from meat, and let him live on bread and milk and farinaceous puddings.

248. *If a child have large bowels, what would you recommend as likely to reduce their size ?*

It ought to be borne in mind, that the bowels of a child are larger in proportion than those of an adult. But, if they be actually larger than they ought to be, let them be well rubbed for a quarter of an hour at a time night and morning, with soap liniment, and then apply a broad flannel belt. " A broad flannel belt worn night and day, firm but not tight, is very serviceable."* The child ought to be prevented from drinking as much as

---

* Sir Charles Locock, in a *Letter* to the Author.

he has been in the habit of doing; let him be encouraged to exercise himself well in the open air; and let strict regard be paid to his diet.

249. *What are the best aperients for a child?*

If it be *actually* necessary to give him opening medicine, one or two tea-spoonfuls of Syrup of Senna, repeated, if necessary, in four hours, will generally answer the purpose; or, for a change, one or two tea-spoonfuls of Castor Oil may be substituted. Lenitive Electuary (Compound Confection of Senna) is another excellent aperient for the young, it being mild in its operation, and pleasant to take; a child fancying it is nothing more than jam, and which it much resembles both in appearance and in taste. The dose is half or one tea-spoonful early in the morning occasionally. Senna is an admirable aperient for a child, and is a safe one, which is more than can be said of many others. It is worthy of note that "the taste of Senna may be concealed by sweeting the infusion,* adding milk, and drinking as ordinary tea, which, when thus prepared, it much resembles."† Honey, too, is a nice aperient for a child—a tea-spoonful ought to be given either by itself, or spread on a slice of bread.

Some mothers are in the habit of giving their children jalap gingerbread. I do not approve of it, as jalap is a drastic, griping purgative; besides, jalap is very nasty to take—nothing will make it palatable.

Fluid Magnesia—Solution of Carbonate of Magnesia—is a good aperient for a child; and, as it has very little taste, is readily given, more especially if made palatable by the addition either of a little syrup or of brown

---

* Infusion of Senna may be procured of any respectable druggist. It will take about one or two table-spoonfuls, or even more, of the infusion (according to the age of the child, and the obstinacy of the bowels), to act as an aperient. Of course, you yourself will be able, from time to time, as the need arises, to add the milk and the sugar, and thus to make it palatable. It ought to be given warm, so as the more to resemble tea.

† *Waring's Manual of Practical Therapeutics.*

sugar. The advantages which it has over the old solid form are, that it is colourless and nearly tasteless, and never forms concretions in the bowels, as the *solid* magnesia, if persevered in for any length of time, sometimes does. A child of two or three years old may take one or two table-spoonfuls of the fluid, either by itself or in his food, repeating it every four hours until the bowels be open. When the child is old enough to drink the draught off *immediately,* the addition of one or two tea-spoonfuls of Lemon Juice to each dose of the Fluid Magnesia, makes a pleasant effervescing draught, and increases its efficacy as an aperient.

Bran-bread* and *treacle* will frequently open the bowels; and as treacle is wholesome, it may be substituted for butter when the bowels are inclined to be costive. A roasted apple, eaten with *raw* sugar, is another excellent mild aperient for a child. Milk gruel—that is to say, milk thickened with oatmeal— forms an excellent food for him, and often keeps his bowels regular, and thus (*which is a very important consideration*) supersedes the necessity of giving him an aperient. An orange (taking care he does not eat the peel or the pulp), or a fig after dinner, or a few Muscatel raisins, will frequently regulate the bowels.

Stewed prunes is another admirable remedy for the costiveness of a child. The manner of stewing them is as follows :—Put a pound of prunes in a brown jar, add two table-spoonfuls of *raw* sugar, then cover the prunes and the sugar with cold water; place them in the oven, and let them stew for four hours. A child should every morning eat half a dozen or a dozen of them, until the bowels be relieved, taking care that he does not swallow the stones. Stewed prunes may be given in treacle— treacle increasing the aperient properties of the prunes.

A suppository is a mild and ready way of opening the bowels of a child. When he is two or three years old

---

* One part of bran to three parts of flour, mixed together and made into bread.

and upwards, a *Candle* suppository is better than a *Soap* suppository. The way of preparing it is as follows :— Cut a piece of dip-tallow candle—the length of three inches—and insert it as you would a clyster pipe, about two inches up the fundament, allowing the remaining inch to be in sight, and there let the suppository remain until the bowels be opened.

Another excellent method of opening a child's bowels is by means of an enema of warm water,—from half a tea-cupful to a tea-cupful, or even more, according to the age of the child. I cannot speak too highly of this plan as a remedy for costiveness, as it entirely, in the generality of cases, prevents the necessity of administering a particle of aperient medicine by the mouth. The fact of its doing so stamps it as a most valuable remedy —opening physic being, as a rule, most objectionable, and injurious to a child's bowels. Bear this fact—for it is a fact—in mind, and let it be always remembered.

450. *What are the most frequent causes of Protrusion of the lower-bowel ?*

The too common and reprehensible practice of a parent · administering frequent aperients, especially calomel and jalap, to her child. Another cause, is allowing him to remain for a quarter of an hour or more at a time on his chair ; this induces him to strain, and to force the gut down.

251. *What are the remedies ?*

If the protrusion of the bowel have been brought on by the abuse of aperients, abstain for the future from giving them ; but if medicine be absolutely required, give the mildest—such as either Syrup of Senna or Castor Oil—*and the less of those the better.*

If the *external* application of a purgative will have the desired effect, it will, in such cases, be better than the *internal* administration of aperients. Castor Oil used as a Liniment is a good one for the purpose. Let the bowels be well rubbed, every night and morning, for five minutes at a time with the oil.

A wet compress to the bowels will frequently open

them, and will thus do away with the necessity of giving an aperient—*a most important consideration.* Fold a napkin in six thicknesses, soak it in *cold* water, and apply it to the bowels ; over which put either a thin covering or sheet of gutta-percha, or a piece of oiled-silk ; keep it in its place with a broad flannel roller ; and let it remain on the bowels for three or four hours, or until they be opened.

Try what diet will do, as opening the bowels by a regulated diet is far preferable to the giving of aperients. Let him have either bran-bread or Robinson's Patent Groats, or Robinson's Pure Scotch Oatmeal made into gruel with new milk, or Du Barry's Arabica Revalenta, or a slice of Huntly and Palmer's lump gingerbread. Let him eat stewed prunes, stewed rhubarb, roasted apples, strawberries, raspberries, the inside of grapes and gooseberries, figs, &c. Give him early every morning a draught of *cold* water.

Let me, again, urge you *not* to give aperients in these cases, or in any case, unless you are absolutely compelled. By following my advice you will save yourself an immense deal of trouble, and your child a long catalogue of misery. Again, I say, look well into the matter, and whenever it be practicable avoid purgatives.

Now, with regard to the best manner of returning the bowel, lay the child upon the bed on his face and bowels, with his hips a little raised ; then smear lard on the forefinger of your right hand (taking care that the nail be cut close), and gently with your fore-finger press the bowel into its proper place. Remember, if the above methods be observed, you cannot do the slightest injury to the bowel ; and the sooner it be returned, the better it will be for the child ; for if the bowel be allowed to remain long down, it may slough or mortify, and death may ensue. The nurse, every time he has a motion, must see that the bowel does not come down, and if it does, she ought instantly to return it. Moreover, the nurse should be careful *not* to allow the child to remain on his chair more than two or three minutes at a time.

Another excellent remedy for the protrusion of the lower bowel, is to use every morning a cold salt and water sitz bath. There need not be more than a depth of three inches of water in the bath ; a small handful of table salt should be dissolved in the water ; a dash of warm water in the winter time must be added, to take off the extreme chill ; and the child ought not to be allowed to sit in the bath for more than one minute, or whilst the mother can count a hundred ; taking care, the while, to throw either a square of flannel or a small shawl over his shoulders. The sitz bath ought to be continued for months, or until the complaint be removed. I cannot speak in too high praise of these baths.

252. *Do you advise me, every spring and fall, to give my child brimstone to purify and sweeten his blood, and as a preventive medicine ?*

Certainly not ; if you wish to take away his appetite, and to weaken and depress him, give brimstone ! Brimstone is not a remedy fit for a child's stomach. The principal use and value of brimstone is as an external application in itch, and as an internal remedy, mixed with other laxatives, in piles—piles being a complaint of adults. In olden times poor unfortunate children were dosed, every spring and fall, with brimstone and treacle to sweeten their blood ! Fortunately for the present race, there is not so much of that folly practised, but still there is room for improvement. To dose a *healthy* child with physic is the grossest absurdity. No, the less physic a delicate child has the better it will be for him, but physic to a healthy child is downright poison ! And brimstone of all medicines ! It is both weakening and depressing to the system, and by opening the pores of the skin and by relaxing the bowels, is likely to give cold, and thus to make a healthy, a sickly child. Sweeten his blood ! It is more likely to weaken his blood, and thus to make his blood impure ! Blood is not made pure by drugs, but by Nature's medicine ; by exercise, by pure air, by wholesome diet, by sleep in a well-ventilated apartment, by regular and thorough ablution. Brimstone a preventive medicine! Preventive medicine—and brimstone especially in the guise of a preventive medicine—is "a mockery, a delusion, and a snare."

253. *When a child is delicate, and his body, without any assignable cause, is gradually wasting away, and the stomach rejects all food that is taken, what plan can be adopted likely to support his strength, and thus probably be the means of saving his life?* .

I have seen, in such a case, great benefit to arise from half a tea-cupful of either strong mutton-broth or of strong beef-tea, used as an enema every four hours.* It should be administered slowly, in order that it may remain in the bowel. If the child be sinking, either a dessert-spoonful of brandy, or half a wine-glassful of port wine, ought to be added to each enema.

The above plan ought only to be adopted if there be *no* diarrhœa. If there be diarrhœa, an enema must *not* be used. Then, provided there be great wasting away, and extreme exhaustion, and other remedies having failed, it would be advisable to give, by the mouth, *raw* beef of the finest quality, which ought to be taken from the hip bone, and should be shredded very fine. All fat and skin must be carefully removed. One or two tea-spoonfuls (according to the age of the child) ought to be given every four hours. The giving of *raw* meat to children in exhaustive diseases, such as excessive long-standing diarrhœa, was introduced into practice by a Russian physician, a Professor Wiesse of St Petersburg. It certainly is, in these cases, a most valuable remedy, and has frequently been the means of snatching such patients from the jaws of death. Children usually take raw meat with avidity and with a relish.

---

* An enema apparatus is an important requisite in every nursery; it may be procured of any respectable surgical instrument maker. The India-rubber Enema Bottle is, for a child's use, a great improvement on the old syringe, as it is not so likely to get out of order, and, moreover, is more easily used.

254. *If a child be naturally delicate, what plan would you recommend to strengthen him?*

I should advise strict attention to the rules above mentioned, and *change of air*—more especially, if it be possible, to the coast. Change of air, sometimes, upon a delicate child, acts like magic, and may restore him to health when all other means have failed. If a girl be delicate, " carry her off to the farm, there to undergo the discipline of new milk, brown bread, early hours, no lessons, and romps in the hay-field."—*Blackwood.* This advice is, of course, equally applicable for a delicate boy, as delicate boys and delicate girls ought to be treated alike. Unfortunately in these very enlightened days ! there is too great a distinction made in the respective management and treatment of boys and girls.

The best medicines for a delicate child will be the wine of iron and cod-liver oil. Give them combined in the manner I shall advise when speaking of the treatment of Rickets.

In diseases of long standing, and that resist the usual remedies, there is nothing like *change of air.* Hippocrates, the father of medicine, says—

"In longis morbis solum mutare."
(In tedious diseases to change the place of residence.)

A child who, in the winter, is always catching cold, whose life during half of the year is one continued catarrh, who is in consequence, likely, if he grow up at all, to grow up a confirmed invalid, ought, during the winter months, to seek another clime ; and if the parents can afford the expense, they should at the beginning of October, cause him to bend his steps to the south of Europe—Mentone being as good a place as they could probably fix upon.

255. *Do you approve of sea bathing for a delicate young child?*

No : he is frequently so frightened by it that the alarm would do him more harm than the bathing would do him good. The better plan would be to have him every

morning well sponged, especially his back and loins, with
sea water ; and to have him as much as possible carried
on the beach, in order that he may inhale the sea breezes.
When he be older, and is not frightened at being dipped,
sea bathing will be very beneficial to him. If bathing
is to do good, either to an adult or to a child, it must be
anticipated with pleasure, and neither with dread nor
with distaste.

256. *What is the best method for administering.medi-
cine to a child?*

If he be old enough, appeal to his reason ; for, if a
mother endeavour to deceive her child, and he detect
her, he will for the future suspect her. If he be too
young to be reasoned with, then, if he will not take his
medicine, he must be compelled. Lay him across your
knees, let both his hands and his nose be tightly held,
and then, by means of the patent medicine-spoon, or, if
that be not at hand, by either a tea or a dessert-spoon,
pour the medicine down his throat, and he will be obliged
to swallow it.

It may be said that this is a cruel procedure ; but it
is the only way to compel an unruly child to take physic,
and is much less cruel than running the risk of his dying
from the medicine not having been administered.*

257. *Ought a sick child to be roused from his sleep to
give him physic, when it is time for him to take it?*

On no account, as sleep, being a natural restorative,
must not be interfered with. A mother cannot be too
particular in administering the medicine, at stated periods,
whilst he is awake.

---

* If any of my medical brethren should perchance read these
Conversations, I respectfully and earnestly recommend them to
take more pains in making medicines for children pleasant and
palatable. I am convinced that, in the generality of instances,
provided a little more care and thought were bestowed on the
subject, it may be done ; and what an amount of both trouble
and annoyance it would save ! It is really painful to witness
the struggles and cries of a child when *nauseous* medicine is to
be given ; the passion and the excitement often do more harm
than the medicine does good

258. *Have you any remarks to make on the manage-ment of a sick-room, and have you any directions to give on the nursing of a child ?*

In sickness select a large and lofty room ; if in the town, the back of the house will be preferable—in order to keep the patient free from noise and bustle—as a sick-chamber cannot be kept too quiet. Be sure that there be a chimney in the room—as there ought to be in *every* room in the house—and that it be not stopped, as it will help to carry off the impure air of the apartment. Keep the chamber *well ventilated*, by, from time to time, open-ing the window. The air of the apartment cannot be too pure ; therefore, let the evacuations from the bowels be instantly removed, either to a distant part of the house, or to an out-house or to the cellar, as it might be necessary to keep them for the medical man's inspection.

Before using either the night-commode, or *the pot-de-chambre*, let a little water, to the depth of one or two inches, he put in the pan, or *pot ;* in order to sweeten the motion, and to prevent the fœcal matter from adhering to the vessel.

Let there be frequent change of linen, as in sickness it is even more necessary than in health, more especially if the complaint be fever. In an attack of fever, clean sheets ought, every other day, to be put on the bed ; clean body-linen every day. A frequent change of linen in sickness is most refreshing.

If the complaint be fever, a fire in the grate will not be necessary. Should it be a case either of inflammation of the lungs or of the chest, a small fire in the winter time is desirable, keeping the temperature of the room as nearly as possible at 60° Fahrenheit. Bear in mind that a large fire in a sick-room cannot be too strongly con-demned ; for if there be fever—and there are scarcely any complaints without—a large fire only increases it. Small fires, in cases either of inflammation of the lungs or of the chest, in the winter time, encourage ventilation of the apartment, and thus carry off impure air. If it be summer time, of course fires would be improper. A

O

thermometer is an indispensable requisite in a sick-room.

In fever, free and thorough ventilation is of vital importance, more especially in scarlet fever; then a patient cannot have too much air; in scarlet fever, for the first few days the windows, be it winter or summer, must to the widest extent be opened. The fear of the patient catching cold by doing so is one of the numerous prejudices and baseless fears that haunt the nursery, and the sooner it is exploded the better it will be for human life. The valances and bed-curtains ought to be removed, and there should be as little furniture in the room as possible.

If it be a case of measles, it will be necessary to adopt a different course; then the windows ought not to be opened, but the door must from time to time be left ajar. In a case of measles, if it be winter time, a *small* fire in the room will be necessary. In inflammation of the lungs or of the chest, the windows should not be opened, but the door ought occasionally to be left unfastened, in order to change the air and to make it pure. Remember, then, that ventilation, either by open window or by open door, is in all diseases most necessary. Ventilation is one of the best friends a doctor has.

In fever, do not load the bed with clothes; in the summer a sheet is sufficient, in winter a sheet and a blanket.

In fever, do not be afraid of allowing the patient plenty either of cold water or of cold toast and water; Nature will tell him when he has had enough. In measles, let the chill be taken off the toast and water.

In *croup*, have always ready a plentiful supply of hot water, in case a warm bath might be required.

In *child-crowing*, have always in the sick-room a supply of cold water, ready at a moment's notice to dash upon the face.

In fever, do not let the little patient lie on the lap; he will rest more comfortably on a horse-hair mattress in his crib or cot. If he have pain in the bowels, the lap is most agreeable to him ; the warmth of the body, either of the mother or of the nurse, soothes him ; besides, if he be on the lap, he can be turned on his stomach and on his bowels, which often affords him great relief and comfort. If he be much emaciated, when he is nursed, place a pillow upon the lap and let him lie upon it.

In *head affections*, darken the room with a *green* calico blind ; keep the chamber more than usually quiet ; let what little talking is necessary be carried on in whispers, but the less of that the better ; and in *head affections*, never allow smelling salts to be applied to the nose, as they only increase the flow of blood to the head, and consequently do harm.

It is often a good sign for a child, who is seriously ill, to suddenly become cross. It is then he begins to feel his weakness and to give vent to his feelings. " Children are almost always cross when recovering from an illness, however patient they may have been during its severest moments, and the phenomenon is not by any means confined to children."—*Geo. M'Donald.*

A sick child must *not* be stuffed with *much* food at a time. He will take either a table-spoonful of new milk or a table-spoonful of chicken broth every half hour with greater advantage than a tea-cupful of either the one or the other every four hours, which large quantity would very probably be rejected from his stomach, and may cause the unfortunately treated child to die of starvation !

If a sick child be peevish, attract his attention either by a toy or by an ornament ; if he be cross, win him over to good humour by love, affection, and caresses, but let it be done gently and without noise. Do not let visitors see him ; they will only excite, distract, and irritate him, and help to consume the oxygen of the atmosphere, and thus rob the air of its exhilarating healthgiving qualities and purity ; a sick-room, therefore, is not a proper place, either for visitors or for gossips.

In selecting a sick-nurse, let her be gentle, patient, cheerful, quiet, and kind, but firm withal ; she ought to be neither old nor young : if she be old she is often garrulous and prejudiced, and thinks too much of her trouble; if she be young, she is frequently thoughtless and noisy; therefore choose a middle-aged woman. Do not let there be in the sick-room more than, besides the mother, one efficient nurse; a greater number can be of no service—they will only be in each other's way, and will distract the patient.

Let stillness, especially if the head be the part affected, reign in a sick-room. Creaking shoes* and rustling silk dresses ought not to be worn in sick-chambers—they are quite out of place there. If the child be asleep, or if he be dozing, perfect stillness must be enjoined, not even a whisper should be heard :—

> " In the sick-room be calm,
>   Move gently and with care,
> Lest any jar or sudden noise,
>   Come sharply unaware.
>
> You cannot tell the harm,
>   The mischief it may bring,
> To wake the sick one suddenly,
>   Besides the suffering.
>
> The broken sleep excites
>   Fresh pain, increased distress ;
> The quiet slumber undisturb'd
>   Soothes pain and restlessness.
>
> Sleep is the gift of God :
>   Oh ! bear these words at heart,
> 'He giveth His beloved sleep,'
>   And gently do thy part." *

If there be other children, let them be removed to a

---

* Nurses at these times ought to wear slippers, and not shoes. The *best* slippers in sick-rooms. are those manufactured by the North British Rubber Company, Edinburgh ; they enable nurses to walk in them about the room without causing the slightest noise ; indeed, they might truly be called " the noiseless slipper," a great desideratum in such cases, more especially in all head affections of children. If the above slippers cannot readily be obtained, then list slippers—soles and all being made of list— will answer the purpose equally as well.

* *Household Verses on Health and Happiness.* London : Jarrold and Sons. .A most delightful little volume.

distant part of the house; or, if the disease be of an infectious nature, let them be sent away from home altogether.

In all illnesses—and bear in mind the following is most important advice—a child must be encouraged to try and make water, whether he ask or not, at least four times during the twenty-four hours; and at any other time, if he express the slightest inclination to do so. I have known a little fellow to hold his water, to his great detriment, for twelve hours, because either the mother had in her trouble forgotten to inquire, or the child himself was either too ill or too indolent to make the attempt.

See that the medical man's directions are, to the very letter, carried out. Do not fancy that you know better than he does, otherwise you have no business to employ him. Let him, then, have your implicit confidence and your exact obedience. What *you* may consider to be a trifling matter, may frequently be of the utmost importance, and may sometimes decide whether the case shall end either in life or death!

Lice.—It is not very poetical, as many of the grim facts of every-day life are not, but, unlike a great deal of poetry, it is unfortunately too true that after a severe and dangerous illness, especially after a bad attack of fever, a child's head frequently becomes infested with vermin —with lice! It therefore behoves a mother herself to thoroughly examine, by means of a fine-tooth comb,* her child's head, in order to satisfy her mind that there be no vermin there. As soon as he be well enough, he ought to resume his regular ablutions—that is to say, that he must go again regularly *into* his tub, and have his head every morning thoroughly washed with soap and water. A mother ought to be particular in seeing

---

* Which fine-tooth comb ought *not* to be used at any other time except for the purpose of examination, as the constant use of a fine-tooth comb would scratch the scalp, and would encourage a quantity of scurf to accumulate

that the nurse washes the hair-brush at least once every week ; **if** she does not do so, the dirty brush which had during the illness been used, might contain the " nits " —the eggs of the lice—and would thus propagate the vermin, as they will, when on the head of the child, soon hatch. If there be already lice on the head, in addition to the regular washing every morning with the soap and water, and after the head has been thoroughly dried, let the hair be well and plentifully dressed with camphorated oil—the oil being allowed to remain on until the next washing on the following morning. Lice cannot live in oil (more especially if, as in camphorated oil, camphor be dissolved in it), and as the camphorated oil will not, in the slightest degree, injure the hair, it is the best application that can be used. But as soon as the vermin have disappeared, let the oil be discontinued, as the *natural* oil of the hair is, at other times, the only oil that is required on the head.

The " nit "—the egg of the louse—might be distinguished from scurf (although to the *naked* eye it is very much like it in appearance). by the former fastening firmly on one of the hairs as a barnacle would on a rock, and by it not being readily brushed off as scurf would, which latter (scurf) is always loose.

259. *My child, in the summer time, is much tormented with fleas : what are the best remedies ?*

A small muslin bag, filled with camphor, placed in the cot or bed, will drive fleas away. Each flea-bite should, from time to time, be dressed by means of a camel's hair brush, with a drop or two of Spirit of Camphor ; an ounce bottle of which ought, for the purpose, to be procured from a chemist. Camphor is also an excellent remedy to prevent bugs from biting. Bugs and fleas have a horror of camphor ; and well they might, for it is death to them !

There is a famous remedy for the destruction of fleas manufactured in France, entitled " *La Poudre Insecticide*," which, although perfectly harmless to the human economy, is utterly destructive to fleas. Bugs are best destroyed either by Creosote or by oil of Turpentine : the places they do love to congregate in should be well saturated by means of a brush, with the creosote or with the oil of turpentine. A few dressings will effectually destroy both them and their young ones.

260. *Is not the pulse a great sign either of health or of disease?*

It is, and every mother should have a general idea of what the pulse of children of different ages should be both in health and in disease. "Every person should know how to ascertain the state of the pulse in health ; then, by comparing it with what it is when he is ailing, he may have some idea of the urgency of his case. Parents should know the healthy pulse of each child, since now and then a person is born with a peculiarly slow or fast pulse, and the very case in hand may be of such peculiarity. An infant's pulse is 140, a child of seven about 80, and from 20 to 60 years it is 70 beats a minute, declining to 60 at fourscore. A healthful grown person beats 70 times in a minute, declining to 60 at fourscore. At 60, if the pulse always exceeds 70, there is a disease ; the machine working itself out, there is a fever or inflammation somewhere, and the body is feeding on itself, as in consumption, when the pulse is quick."

261. *Suppose a child to have had an attack either of inflammation of the lungs or of bronchitis, and to be much predisposed to a return : what precautions would you take to prevent either the one or the other for the future?*

I would recommend him to wear fine flannel instead of lawn shirts ; to wear good lamb's-wool stockings *above the knees*, and good, strong, dry shoes to his feet ; to live, weather permitting, a great part of every day in the open air ; to strengthen his system by good nourishing food —by an abundance of both milk and meat (the former especially) ; to send him, in the autumn, for a couple of months, to the sea-side ; to administer to him, from time to time, cod-liver oil ; in short, to think only of his

health, and to let learning, until he be stronger, be left
alone.   I also advise either table salt or bay salt, or Tidman's Sea Salt, to be added to the water in which the
child is washed with in the morning, in a similar manner
as recommended in answer to a previous question.

262. *Then do you not advise such a child to be confined within doors ?*

If any inflammation be present, or if he have but just
recovered from one, it would be improper to send him
into the open air, but not otherwise, as the fresh air
would be a likely means of strengthening the lungs, and
thereby of preventing an attack of inflammation for the
future.   Besides, the more a child is coddled within
doors, the more likely will he be to catch cold, and to
renew the inflammation.   If the weather be cold, yet
neither wet nor damp, he ought to be sent out, but let
him be well clothed ; and the nurse should have strict
injunctions *not* to stand about entries or in any draughts
—indeed, not to stand about at all, but to keep walking
about all the time she is in the open air.   Unless you
have a trustworthy nurse, it will be well for you either
to accompany her in her walk with your child, or merely
to allow her to walk with him in the garden, as you can
then keep your eye upon both of them.

263. *If a child be either chicken-breasted, or if he be
narrow-chested, are there any means of expanding and of
strengthening his chest ?*

Learning ought to be put out of the question, attention must be paid to his health alone, or consumption
will probably mark him as its own !   Let him live as
much as possible in the open air ; if it be country, so
much the better.   Let him rise early in the morning,
and let him go to bed betimes ; and if he be old enough
to use the dumb-bells, or what is better, an India-rubber
chest-expander, he should do so daily.   He ought also
to be encouraged to use two short sticks, similar to,
but heavier than, a policeman's staff, and to go, every
morning, through regular exercises with them.   As soon
as he is old enough, let him have lessons from a drill-

sergeant and from a dancing-master. Let him be made both to walk and to sit upright, and let him be kept as much as possible upon a milk diet,* and give him as much as he can eat of fresh meat every day. Cod-liver oil, a tea-spoonful or a dessert-spoonful, according to his age, twice a day, is serviceable in these cases. Stimulants ought to be carefully avoided. In short, let every means be used to nourish, to strengthen, and invigorate the system, without, at the same time, creating fever. Such a child should be a child of nature; he ought almost to live in the open air, and throw his books to the winds. Of what use is learning without health? In such a case as this you cannot have both.

264. *If a child be round-shouldered, or if either of his shoulder-blades have " grown out," what had better be done ?*

Many children have either round-shoulders, or have their shoulder-blades grown out, or have their spines twisted, from growing too fast, from being allowed to slouch in their gait, and from not having sufficient nourishing food, such as meat and milk, to support them while the rapid growth of childhood is going on.

If your child be affected as above described, nourish him well on milk and on farinaceous food, and on meat once a day, but let milk be his staple diet; he ought, during the twenty-four hours, to take two or three pints of new milk. He should almost live in the open air, and must have plenty of play. If you can so contrive it, let him live in the country. When tired, let him lie, for half an hour, two or three times daily, flat on his back on the carpet. Let him rest at night on a horse-hair mattress, and not on a feather bed.

Let him have every morning, if it be summer, a thorough cold water ablution ; if it be winter, let the

---

* Where milk does not agree, it may generally be made to do so by the addition of one part of lime water to seven parts of new milk. Moreover, the lime will be of service in hardening his bones ; and, in these cases, the bones require hardening.

water be made tepid. Let either two handfuls of table salt or a handful of bay salt be dissolved in the water. Let the salt and water stream well over his shoulders and down his back and loins. Let him be well dried with a moderately coarse towel, and then let his back be well rubbed, and his shoulders be thrown back—exercising them much in the same manner as in skipping, for five or ten minutes at a time. Skipping, by-the-by, is of great use in these cases, whether the child be either a boy or a girl—using, of course, the rope backwards, and not forwards.

Let books be utterly discarded until his shoulders have become strong, and thus no longer round, and his shoulder-blades have become straight. It is a painful sight to see a child stoop like an old man.

Let him have, twice daily, a tea-spoonful or a dessert-spoonful (according to his age) of cod-liver oil, giving it him on a full and not on an empty stomach.

When he is old enough, let the drill-sergeant give him regular lessons, and let the dancing-master be put in requisition. Let him go through regular gymnastic exercises, provided they are not of a violent character.

But, bear in mind, let there be in these cases no mechanical restraints—no shoulder-straps, no abomin- able stays. Make him straight by natural means—by making him strong. Mechanical means would only, by weakening and wasting the muscles, increase the mischief, and thus the deformity. In this world of ours there is too much reliance placed on artificial, and too little on natural means of cure.

265. *What are the causes of Bow Legs in a child ; and what is the treatment ?*

Weakness of constitution, poor and insufficient nourishment, and putting a child, more especially a fat and heavy one, on his legs too early.

Treatment.—Nourishing food, such as an abundance of milk, and, if he be old enough, of meat ; iron medicines ; cod-liver oil ; thorough ablution, every morning, of the whole body ; an abundance of exercise, either on pony, or on donkey, or in carriage, but not, until his legs be stronger, on foot. If they are much bowed, it will be necessary to consult an experienced surgeon.

266. If a child, while asleep, " wet his bed," is there any method of preventing him from doing so ?

Let him be held out just before he himself goes to bed, and again when the family retires to rest. If, at the time, he be asleep, he will become so accustomed to it, that he will, without awaking, make water. He ought to be made to lie on his side ; for, if he be put on his back, the urine will rest upon an irritable part of the bladder, and, if he be inclined to wet his bed, he will not be able to avoid doing so. He must not be allowed to drink much with his meals, especially with his supper. Wetting the bed is an infirmity with some children—they cannot help it. It is, therefore, cruel to scold and chastise them for it. Occasionally, however, wetting the bed arises from idleness ; in which case, of course, a little wholesome correction might be neces- sary.

Water-proof Bed-sheeting—one yard by three-quarters of a yard—will effectually preserve the bed from being wetted, and ought always, on these occasions, to be used.

A mother ought, every morning, to ascertain for her- self, whether a child have wet his bed ; if he have, and if, unfortunately, the water-proof cloth have not been used, the mattress, sheets, and blankets must be instantly taken to the kitchen fire and be properly dried. Inatten- tion to the above has frequently caused a child to suffer either from cold, from a fever, or from an inflamma- tion ; not only so, but, if they be not dried, he is wallow- ing in filth and in an offensive effluvium. If both mother and nurse were more attentive to their duties— in frequently holding a child out, whether he ask or not—a child wetting his bed would be the exception, and not, as it frequently is, the rule. If a child be dirty, you may depend upon it, the right persons to blame are the mother and the nurse, and not the child !

267. *If a child should catch Small-pox, what are the best means to prevent pitting ?*

He ought to be desired neither to pick nor to rub the pustules. If he be too young to attend to these directions, his hands must be secured in bags (just large enough to hold them), which bags should be fastened round the wrists. The nails must be cut very close.

Cream smeared, by means of a feather, frequently in the day, on the pustules, affords great comfort and benefit. Tripe liquor (without salt) has, for the· same purpose, been strongly recommended. I myself, in several cases, have tried it, and with the happiest results. It is most soothing, comforting, and healing to the skin.

268. *Can you tell me of any plan to prevent Chilblains, or, if a child be suffering from them, to cure them ?*

First, then, the way to prevent them.—Let a child, who is subject to them, wear, in the winter time, a square piece of wash-leather over the toes, a pair of warm lamb's-wool stockings, and good shoes; but, above all, let him be encouraged to run about the house as much as possible, especially before going to bed ; and on no account allow him either to warm his feet before the fire, or to bathe them in *hot* water. If the feet be cold, and the child be too young to take exercise, then let them be well rubbed with the warm hand. If adults suffer from chilblains, I have found friction, night and morning, with horse-hair flesh-gloves, the best means of preventing them.

Secondly, the way to cure them.—*If they be unbroken :* the old-fashioned remedy of onion and salt is one of the best of remedies. Cut an onion in two ; take one-half of it, dip it in table salt and well rub, for two or three minutes, the chilblain with it. The onion and salt is a famous remedy to relieve that intolerable itching which sometimes accompanies chilblains : then let them be covered with a piece of lint, over which a piece of wash-leather should be placed. .

If they be broken, let a piece of lint be spread with spermaceti-cerate, and be applied, every morning, to the part, and let a white-bread poultice be used every night.

269. *During the winter time my child's hands, legs, &c., chap very much ; what ought I to do ?*

Let a tea-cupful of bran be tied up in a muslin bag, and be put, over the night, into either a large water-can or jug of *rain* water; * and let this water from the can or jug be the water he is to be washed with on the following morning, and every morning until the chaps be cured. As often as water is withdrawn, either from the water-can or from the jug, let fresh rain water take its place, in order that the bran may be constantly soaking in it. The bran in the bag should be renewed about twice a week.

Take particular care to dry the skin well every time he be washed; then, after each ablution, as well as every night at bed-time, rub a piece of deer's suet over the parts affected : a few dressings will perform a cure. The deer's suet may be bought at any of the shops where venison is sold. Another excellent remedy is glycerine,† which should be smeared, by means of the finger or by a camel's hair brush, on the parts affected, two or three times a day. If the child be very young, it might be necessary to dilute the glycerine with rose-water ; fill a small bottle one-third with glycerine, and fill up the remaining two-thirds of the bottle with rose-water— shaking the bottle every time just before using it. The best soap to use for chapped hands is the glycerine soap : no other being required.

270. *What is the best remedy for Chapped Lips ?*

Cold-cream (which may be procured of any respect-

---

* *Rain* water ought *always* to be used in the washing of a child ; pump water is likely to chap the skin, and to make it both rough and irritable.

† Glycerine prepared by Price's Patent Candle Company is by far the best. Sometimes, if the child's skin be very irritable, the glycerine requires diluting with water—say, two ounces of glycerine to be mixed in a bottle with four ounces of rain water—the bottle to be well shaken just before using it.

able chemist) is an excellent application for *chapped lips*. It ought, by means of the finger, to be frequently smeared on the parts affected.

271. *Have the goodness to inform me of the different varieties of Worms that infest a child's bowels ?*

Principally three—1, The tape-worm ; 2, the long round-worm ; and 3, the most frequent of all, the common thread or maw-worm. The tape-worm infests the whole course of the bowels, both small and large : the long round-worm, principally the small bowels, occasionally the stomach ; it sometimes crawls out of the child's mouth, causing alarm to the mother ; there is, of course, no danger in its doing so : the common thread-worm or maw-worm infests the rectum or fundament.

272. *What are the causes of Worms ?*

The causes of worms are : weak bowels ; bad and improper food, such as unripe, unsound, or uncooked fruit, and much green vegetables ; pork, especially underdone pork ; * an abundance of sweets ; the neglecting of giving salt in the food.

273. *What are the symptoms and the treatment of Worms ?*

The symptoms of worms are—emaciation ; itching and picking of the nose ; a dark mark under the eyes ; grating, during sleep, of the teeth ; starting in the sleep ; foul breath ; furred tongue ; uncertain appetite—sometimes voracious, at other times bad, the little patient sitting down very hungry to his dinner, and before scarcely tasting a mouthful, the appetite vanishing ; large bowels ; colicky pains of the bowels ; slimy motions ; itching of the fundament. Tape-worm and round-worm, more especially the former, are apt, in children,

---

* One frequent, if not the most frequent, cause of tape-worm is the eating of pork, more especially if it be underdone. *Underdone* pork is the most unwholesome food that can be eaten, and is the most frequent cause of tape-worm known. *Underdone* beef also gives tape-worm ; let the meat, therefore, be well and properly cooked. These facts ought to be borne in mind, as prevention is always better than cure.

to produce convulsions. Tape-worm is very weakening to the constitution, and usually causes great emaciation and general ill-health; the sooner, therefore, it is expelled from the bowels the better it will be for the patient.

Many of the obscure diseases of children arise from worms. In all doubtful cases, therefore, this fact should be borne in mind, in order that a thorough investigation may be instituted.

With regard to *treatment,* a medical man ought, of course, to be consulted. He will soon use means both to dislodge them, and to prevent a future recurrence of them.

Let me caution a mother never to give her child patent medicines for the destruction of worms. There is one favourite quack powder, which is composed principally of large doses of calomel, and which is quite as likely to destroy the patient as the worms! No, if your child have worms, put him under the care of a judicious medical man, who will soon expel them, without, at the same time, injuring health or constitution!

274. *How may worms be prevented from infesting a child's bowels?*

Worms generally infest *weak* bowels; hence, the moment a child becomes strong worms cease to exist. The reason why a child is so subject to them is owing to the improper food which is usually given to him. When he be stuffed with unsound and with unripe fruits, with much sweets, with rich puddings, and with pastry, and when he is oftentimes allowed to eat his meat *without* salt, and to *bolt* his food without chewing it, is there any wonder that he should suffer from worms? The way to prevent them is to avoid such things, and, at the same time, to give him plenty of salt to his *fresh* and well-cooked meat. Salt strengthens and assists digestion, and is absolutely necessary to the human economy. Salt is emphatically a worm destroyer. The truth of this statement may be readily tested by

sprinkling a little salt on the common earth-worm.
" What a comfort and real requisite to human life is
salt ! It enters into the constituents of the human
blood, and to do without it is wholly impossible."—*The
Grocer.* To do without it is wholly impossible ! These
are true words. Look well to it, therefore, ye mothers,
and beware of the consequences of neglecting such
advice, and see for yourselves that your children
regularly eat salt with their food. If they neglect eating
salt with their food, they *must of necessity have worms,*
and worms that will eventually injure them, and make
them miserable. All food, then, should be " flavoured
with salt ;" *flavoured,* that is to say, salt should be used
in each and every kind of food—*not in excess, but in
moderation.*

275. *You have a great objection to the frequent administration of aperient medicines to a child : can you
advise any method to prevent their use ?*

Although we can scarcely call constipation a disease,
yet it sometimes leads to disease. The frequent giving
of aperients only adds to the stubbornness of the
bowels.

I have generally found a draught, early every morning,
of *cold* pump water, the eating either of Huntley and
Palmer's loaf ginger-bread, or of oatmeal gingerbread, a
variety of animal and vegetable food, ripe sound fruit,
Muscatel raisins, a fig, or an orange .after dinner, and,
when he be old enough, *coffee* and milk instead of *tea*
and milk, to have the desired effect, more especially if,
for a time, aperients be studiously avoided.

276. *Have you any remarks to make on Rickets ?*

Rickets is owing to a want of a sufficient quantity of
earthy matter in the bones ; hence the bones bend and
twist, and lose their shape, causing deformity. Rickets
generally begins to show itself between the first and
second years of a child's life. Such children are
generally late in cutting their teeth, and when the teeth
do come they are bad, deficient of enamel, discoloured,
and readily decay. A rickety child is generally stunted

in stature ; he has a large head, with overhanging fore-head, or what nurses call a watery-head-shaped forehead. The fontanelles, or openings of the head, as they are called, are a long time in closing. A rickety child is usually talented ; his brain seems to thrive at the expense of his general health. His breast-bone projects out, and the sides of his chest are flattened ; hence he becomes what is called chicken-breasted or pigeon-breasted ; his spine is usually twisted, so that he is quite awry, and, in a bad case, he is hump-backed ; the ribs, from the twisted spine, on one side bulge out ; he is round-shouldered ; the long bones of his body, being soft, bend ; he is bow-legged, knock-kneed, and weak-ankled.

Rickets are of various degrees of intensity, the hump-backed being among the worst. There are many mild forms of rickets ; weak ankles, knocked-knees, bowed-legs, chicken-breasts, being among the latter number. Many a child, who is not exactly hump-backed, is very round-shouldered, which latter is also a mild species of rickets.

Show me a child that is rickety, and I can generally prove that it is owing to poor living, more especially to poor milk. If milk were always genuine, and if a child had an abundance of it, my belief is that rickets would be a very rare disease. The importance of genuine milk is of national importance. We cannot have a race of strong men and women unless, as children, they have had a good and plentiful supply of milk. It is utterly impossible. Milk might well be considered one of the necessaries of a child's existence. Genuine, fresh milk, then, is one of the grand preventatives, as well as one of the best remedies, for rickets. Many a child would not now have to swallow quantities of cod-liver oil if previously he had imbibed quantities of good genuine milk. An insufficient and a poor supply of milk in childhood sows the seeds of many diseases, and death often gathers the fruit. Can it be wondered at, when there is so much poor and nasty milk in England, that rickets in one shape or another is so prevalent ?

When will mothers arouse from their slumbers, rub their eyes, and see clearly the importance of the subject? When will they know that all the symptoms of rickets I have just enumerated *usually* proceed from the want of nourishment, more especially from the want of genuine, and of an abundance of, milk? There are, of, course, other means of warding off rickets besides an abundance of nourishing food, such as thorough ablution, plenty of air, exercise, play, and sunshine · but of all these splendid remedies, nourishment stands at the top of the list.

I do not mean to say that rickets *always* proceeds from poorness of living—from poor milk. It sometimes arises from scrofula, and is an inheritance of one or of both the parents.

Rickety children, if not both carefully watched and managed, frequently, when they become youths, die of consumption.

, A mother, who has for some time neglected the advice I have just given, will often find, to her grievous cost, that the mischief has, past remedy, been done, and that it is now "too late!—too late!"

277. *How may a child be prevented from becoming rickety? or, if he be rickety, how ought he to be treated?*

If a child be predisposed to be rickety, or if he be actually rickety, attend to the following rules :—

Let him live well, on good nourishing diet, such as on tender rump-steaks, cut very fine, and mixed with mashed potatoes, crumb of bread, and with the gravy of the meat. Let him have, as I have before advised, an abundance of good new milk—a quart or three pints during every twenty-four hours. Let him have milk in every form—as milk gruel, Du Barry's Arabica Revalenta made with milk, batter and rice puddings, suet puddings, bread and milk, &c.

To harden the bones, let lime water be added to the milk (a table-spoonful to each tea-cupful of milk.)

Let him have a good supply of fresh, pure, dry air. He must almost live in the open air—the country, if practicable, in preference to the town, and the coast in summer and autumn. Sea bathing and sea breezes are often, in these cases, of inestimable value.

He ought not, at an early age, to be allowed to bear his weight upon his legs. He must sleep on a horse-hair mattress, and not on a feather bed. He should use every morning cold baths in the summer and tepid baths in the winter, with bay salt (a handful) dissolved in the water.

Friction with the hand must, for half an hour at a time, every night and morning, be sedulously applied to the back and to the limbs. It is wonderful how much good in these cases friction does.

Strict attention ought to be paid to the rules of health as laid down in these Conversations. Whatever is conducive to the general health is preventive and curative of rickets.

Books, if he be old enough to read them, should be thrown aside; health, and health alone, must be the one grand object.

The best medicines in these cases are a combination of cod-liver oil and the wine of iron, given in the following manner:—Put a tea-spoonful of wine of iron into a wine-glass, half fill the glass with water, sweeten it with a lump or two of sugar, then let a tea-spoonful of cod-liver oil swim on the top; let the child drink it all down together, twice or three times a day. An hour after a meal is the *best* time to give the medicine, as both iron and cod-liver oil sit better on a *full* than on an *empty* stomach. The child in a short time will become fond of the above medicine, and will be sorry when it is discontinued.

A case of rickets requires great patience and steady perseverance; let, therefore, the above plan have a fair and long-continued trial, and I can then promise that there will be every probability that great benefit will be derived from it.

278. *If a child be subject to a scabby eruption about the mouth, what is the best local application?*

Leave it to nature. Do not, on any account, apply any local application to heal it; if you do, you may produce injury; you may either bring on an attack of inflammation, or you may throw him into convulsions. No! This "breaking-out" is frequently a safety-valve, and must not therefore be needlessly interfered with. Should the eruption be severe, reduce the child's diet; keep him from butter, from gravy, and from fat meat, or, indeed, for a few days from meat altogether ; and give him mild aperient medicine; but, above all things, do not quack him either with calomel or with grey-powder.

279 *Will you have the goodness to describe the eruption on the face and on the head of a young child, called Milk-Crust or Running Scall ?*

Milk-crust is a complaint of very young children—of those who are cutting their teeth—and, as it is a nasty looking complaint, and frequently gives a mother a great deal of trouble, of anxiety, and annoyance, it will be well that you should know its symptoms, its causes, and its probable duration.

Symptoms.—When a child is about nine months or a year old, small pimples are apt to break out around the ears, on the forehead, and on the head. These pimples at length become vesicles (that is to say, they contain water), which run into one large one, break, and form a nasty dirty-looking yellowish, and sometimes greenish, scab, which scab is moist, indeed, sometimes quite wet, and gives out a disagreeable odour, and which is sometimes so large on the head as actually to form a skullcap, and so extensive on the face as to form a mask ! These, I am happy to say, are rare cases. The child's beauty is, of course, for a time completely destroyed, and not only his beauty, but his good temper; for as the eruption causes great irritation and itching, he is constantly clawing himself, and crying with annoyance the great part of the day, and sometimes also of the night—the eruption preventing him from sleeping. It is not contagious, and soon after he has cut the whole of his *first* set of teeth it will get well, provided it has not been improperly interfered with.

Causes.—Irritation from teething; stuffing him with overmuch meat, thus producing a humour, which Nature tries to get rid of by throwing it out on the surface of the body; the safest place she could fix on for the purpose; hence the folly and danger of giving medicines and applying *external* applications to drive the eruption in. "Diseased nature oftentimes breaks forth in strange eruptions," and cures herself in this way, if she be not too much interfered with, and if the eruption be not driven in by injudicious treatment. I have known in such cases disastrous consequences to follow over-officiousness and meddlesomeness. Nature is trying all she can to drive the humour out, while some wiseacres are doing all they can to drive the humour in.

Duration.—As milk-crust is a tedious affair, and will require a variety of treatment, it will be necessary to consult an experienced medical man; and although he will be able to afford great relief, the child will not, in all probability, be quite free from the eruption until he have cut the whole of his first set of teeth—until he be upwards of two years and a half old—when, with judicious and careful treatment, it will gradually disappear, and eventually leave not a trace behind.

It will be far better to leave the case alone—to get well of itself—rather than to try to cure the complaint either by outward applications or by strong internal medicines; "the remedy is often worse than the disease," of this I am quite convinced.

280. *Have you any advice to give me as to my conduct towards my medical man* ·

Give him your entire confidence. Be truthful and be candid with him. Tell him the truth, the whole truth, and nothing but the truth. Have no reservations; give him, as near as you can, a plain, unvarnished statement of the symptoms of the disease. Do not magnify, and do not make too light of any of them. Be prepared to state the exact time the child first showed symptoms of illness. If he have had a shivering fit, however slight, do not fail to tell your medical man of it. Note the state of the skin; if there be a "breaking-out"—be it ever so trifling—let it be pointed out to him. Make yourself acquainted with the quantity and with the appearance of the urine, taking care to have a little of it saved, in case the doctor may wish to see and examine it. Take notice of the state of the motions—their number during the twenty-four hours, their colour, their smell, and their consistence, keeping one for his inspection. Never leave any of these questions to be answered by a servant; a mother is the proper person to give the necessary and truthful answers, which answers frequently decide the fate of the patient. Bear in mind, then, a mother's untiring care and love, attention and truthfulness, frequently decide whether, in a serious illness, the little fellow shall live or die! Fearful responsibility!

A medical man has arduous duties to perform; smooth, therefore, his path as much as you can, and you will be amply repaid by the increased good he will be able to do your child. Strictly obey a doctor's orders—in diet, in medicine, in everything. Never throw obstacles in his way. Never omit any of his suggestions; for, depend upon it that if he be a sensible man, directions, however slight, ought never to be neglected; bear in mind, with a judicious medical man,

"That nothing walks with aimless feet."—*Tennyson.*

If the case be severe, requiring a second opinion, never of your own accord call in a physician, without first consulting and advising with your own medical man. It would be an act of great discourtesy to do so. Inattention to the foregoing advice has frequently caused injury to the patient, and heart-burnings and ill-will among doctors.

Speak, in the presence of your child, with respect and kindness of your medical man, so that the former may look upon the latter as a friend—as one who will strive, with God's blessing, to relieve his pain and suffering. Remember the increased power of doing good the doctor will have if the child be induced to like, instead of dislike, him. Not only be careful that you yourself speak before your child respectfully and kindly of the medical man, but see that your domestics do so likewise; and take care that they are never allowed to frighten your child, as many silly servants do, by saying that they will send for the doctor, who will either give him nasty medicine, or will perform some cruel operation upon him. A nurse-maid should, then, never for one moment be permitted to make a doctor an object of terror or of dislike to a child.

Send, whenever it be practicable, for your doctor *early* in the morning, as he will then make his arrangements accordingly, and can by daylight better ascertain the nature of the complaint, more especially if it be a skin disease. It is utterly impossible for him to form a correct opinion of the nature of a "breaking-out" either by gas or by candle light. If the illness come on at night, particularly if it be ushered in either with a severe shivering, or with any other urgent symptom, no time should be lost, be it night or day, in sending for him.

> " A little fire is quickly trodden out ,
> Which, being suffer'd, rivers cannot quench."
> *Shakspeare.*

WARM BATHS.

281. *Have the goodness to mention the complaints of a child for which warm baths are useful.*

1. Convulsions; 2. Pains in the bowels, known by the child drawing up his legs, screaming violently, &c.; 3. Restlessness from teething; 4. Flatulence. The warm bath acts as a fomentation to the stomach and the bowels, and gives ease where the usual remedies do not rapidly relieve.

282. *Will you mention the precautions, and the rules to be observed in putting a child into a warm bath?*

Carefully ascertain before he be immersed in the bath that the water be neither too hot nor too cold. Carelessness, or over-anxiety to put him in the water as quickly as possible, has frequently, from his being immersed in the bath when the water was too hot, caused him great pain and suffering. From 96 to 98 degrees of Fahrenheit is the proper temperature of a warm bath. If it be necessary to add fresh warm water, let him be either removed the while, or let it not be put in when very hot; for if boiling water be added to increase the heat of the bath, it naturally ascends, and may scald him. Again, let the fresh water be put in at as great a distance from him as possible. The usual time for him to remain in a bath is a quarter of an hour or twenty minutes. Let the chest and the bowels be rubbed with the hand while he is in the bath. Let him be immersed in the bath as high up as the neck, taking care that he be the while supported under the armpits, and that his head be also rested. As soon as he comes out of the bath, he ought to be carefully but quickly rubbed dry; and if it be necessary to keep up the action on the skin, he should be put to bed, between the blankets; or if the desired relief has been obtained, between the sheets, which ought to have been previously warmed, where, most likely, he will fall into a sweet refreshing sleep.

WARM EXTERNAL APPLICATIONS.

283. *In case of a child suffering pain either in his stomach or in his bowels, or in case he has a feverish cold, can you tell me of the best way of applying heat to them?*

In pain either of the stomach or of the bowels, there is nothing usually affords greater or speedier relief than the *external* application of heat. The following are four different methods of applying heat:—1. A bag of hot salt—that is to say, powdered table-salt—put either into the oven or into a frying-pan over the fire, and thus made hot, and placed in a flannel bag, and then applied, as the case may be, either to the stomach or to the bowels. Hot salt is an excellent remedy for these pains. 2. An india-rubber hot-water bottle,* half filled with hot water—it need not be boiling—applied to the stomach or to the bowels, will afford great comfort. 3. Another and an excellent remedy for these cases is a hot bran poultice. The way to make it is as follows :—Stir bran into a vessel containing either a pint or a quart (according to size of poultice required) of boiling water, until it be of the consistence of a nice soft poultice, then put into a flannel bag and apply it to the part affected. When cool, dip it from time to time in *hot* water. 4. In case a child has a feverish cold, especially if it be attended, as it sometimes is, with pains in the bowels, the following is a good external application :—Take a yard of flannel, fold it in three widths, then dip it in very hot water, wring it out tolerably dry, and apply it evenly and neatly round and round the bowels; over this, and to keep it in its place, and to keep in the moisture, put on a *dry* flannel bandage, four yards long and four inches wide. If it be put on at bed-time, it ought to remain on all night. Where there are children, it is desirable to have the yard of flannel and the flannel bandage in readiness, and then a mother will be prepared for emergencies. Either the one or the other, then, of the above applications will usually, in pains of the stomach and bowels, afford great relief. There is one great advantage of the *external* application of heat—it can never do harm; if there be inflammation, it will do good; if there be either cramps or spasms of the stomach, it will be serviceable; if there be colic, it will be one of the best remedies that can be used; if it be a feverish cold, by throwing the child into a perspiration, it will be beneficial.

It is well for a mother to know how to make a white bread poultice; and as the celebrated Abernethy was noted for his poultices, I will give you his directions,

---

* Every house where there are children ought to have one of these India-rubber hot-water bottles. It may be procured at any respectable Vulcanised India-rubber warehouse.

and in his very words :—"Scald out a basin, for you can never make a good poultice unless you have perfectly boiling water, then, having put in some hot water, throw in coarsely crumbled bread, and cover it with a plate. When the bread has soaked up as much water as it will imbibe, drain off the remaining water, and there will be left a light pulp. Spread it a third of an inch thick on folded linen, and apply it when of the temperature of a warm bath. It may be said that this poultice will be very inconvenient if there be no lard in it, for it will soon get dry; but this is the very thing you want, and it can easily be moistened by dropping warm water on it, whilst a greasy poultice will be moist, but not wet." —*South's Household Surgery.*

ACCIDENTS.

284. *Supposing a child to cut his finger, what is the best application ?*

There is nothing better than tying it up with rag in its blood, as nothing is more healing than blood. Do not wash the blood away, but apply the rag at once, taking care that no foreign substance be left in the wound. If there be either glass or dirt in it, it will of course be necessary to bathe the cut in warm water, to get rid of it before the rag be applied. Some mothers use either salt or Fryar's Balsam, or turpentine, to a fresh wound; these plans are cruel and unnecessary, and frequently make the cut difficult to heal. If it bleed immoderately, sponge the wound freely with cold water. If it be a severe cut, surgical aid, of course, will be required.

285. *If a child receive a blow, causing a bruise, what had better be done ?*

Immediately smear a small lump of *fresh* butter on the part affected, and renew it every few minutes for two or three hours; this is an old-fashioned, but a very good remedy. Olive oil may—if *fresh* butter be not at hand—be used, or soak a piece of brown-paper in one third of French brandy and two-thirds of water, and immediately apply it to the part; when dry renew it,

Either of these simple plans—the butter plan is the best—will generally prevent both swelling and disfiguration.

A "*Black Eye.*"—If a child, or indeed any one else, receive a blow over the eye, which is likely to cause a "black eye," there is no remedy superior to, nor more likely to prevent one, than well buttering the parts for two or three inches around the eye with fresh butter, renewing it every few minutes for the space of an hour or two; if such be well and perseveringly done, the disagreeable appearance of a "black eye" will in all probability be prevented. A capital remedy for a "black eye" is the Arnica Lotion,—

> Take of—Tincture of Arnica, one ounce ;
> Water, seven ounces ;
> To make a Lotion. The eye to be bathed by means of a soft piece of linen rag, with this lotion frequently ; and, between times, let a piece of linen rag, wetted in the lotion, be applied to the eye, and be fastened in its place by means of a bandage.

The white lily leaf, soaked in brandy, is another excellent remedy for the bruises of a child. Gather the white lily blossoms when in full bloom, and put them in a wide-mouthed bottle of brandy, cork the bottle, and it will then always be ready for use. Apply a leaf to the part affected, and bind it on either with a bandage or with a handkerchief. The white lily root sliced is another valuable external application for bruises.

286. *If a child fall upon his head and be stunned, what ought to be done?*

If he fall upon his head and be stunned, he will look deadly pale, very much as if he had fainted. He will in a few minutes, in all probability, regain his consciousness. Sickness frequently supervenes, which makes the case more serious, it being a proof that injury, more or less severe, has been done to the brain ; send, therefore, instantly for a medical man.

In the meantime, loosen both his collar and neckerchief, lay him flat on his back, sprinkle cold water upon his face, open the windows so as to admit plenty of fresh air, and do not let people crowd round him, nor shout at him, as some do, to make him speak.

While he is in an unconscious state, do not on any account whatever allow a drop of blood to be taken from him, either by leeches or from the arm—venesection ; if you do, he will probably never rally, but will most likely " sleep the sleep that knows not breaking."

287. *A nurse sometimes drops an infant and injures his back ; what ought to be done ?*

Instantly send for a surgeon ; omitting to have proper advice in such a case has frequently made a child a cripple for life. A nurse frequently, when she has dropped her little charge, is afraid to tell her mistress ; the consequences might then be deplorable. If ever a child scream violently without any assignable cause, and the mother is not able for some time to pacify him, the safer plan is that she send for a doctor, in order that he might strip and carefully examine him ; much after misery might often be averted if this plan were more frequently followed.

288. *Have you any remarks to make and directions to give on accidental poisoning by lotions, by liniments, &c. ?*

It is a culpable practice of either a mother or nurse to leave *external* applications within the reach of a child. It is also highly improper to put a mixture and an *external* application (such as a lotion or a liniment) on the same tray or on the same mantel-piece. Many liniments contain large quantities of opium, a tea-spoonful of which would be likely to cause the death of a child. " Hartshorn and oil," too, has frequently been swallowed by children, and in several instances has caused death. Many lotions contain sugar of lead, which is also poisonous. There is not, fortunately, generally sufficient lead in the lotion to cause death ; but if there be not enough to cause death, there may be more than enough to make the child very poorly. All these accidents occur from disgraceful carelessness.

A mother or a nurse ought *always*, before administer- .ing a dose of medicine to a child, to read the label on the bottle ; by adopting this simple plan many serious accidents and much after misery might be averted. . Again, I say, let every lotion, every liniment, and indeed everything for *external* use, be either locked up or be put out of the way, and far away from all medicine that is given by the mouth. *This advice admits of no exception.*

If your child have swallowed a portion of a liniment containing opium, instantly send for a medical man. In the meantime force a strong mustard emetic (composed of two tea-spoonfuls of flour of mustard, mixed in half a tea-cupful of warm water) down his throat. Encourage the vomiting by afterwards forcing him to swallow warm water. Tickle the throat either with your finger or with a feather. Souse him alternately in hot and then in a cold bath. Dash cold water on his head and face. Throw open the windows. Walk him about in the open air. Rouse him by slapping him, by pinching him, and by shouting to him ; rouse him, indeed, by every means in your power, for if you allow him to go to sleep, it will, in all probability, be the sleep that knows no waking !

If a child have swallowed "hartshorn and oil," force him to drink vinegar and water, lemon-juice and water sweetened with sugar, barley water, and thin gruel.

If he have swallowed a lead lotion, give him a mustard emetic, and then vinegar and water, sweetened either with honey or with sugar, to drink.

289. *Are not lucifer matches poisonous ?*

Certainly, they are very poisonous ; it is, therefore, desirable that they should be put out of the reach of children. A mother ought to be very strict with servants on this head. Moreover, lucifer matches are not only poisonous but dangerous, as a child might set himself on fire with them. A case bearing on the subject has just come under my own observation. A little boy three years old, was left alone for two or three minutes, during which time he obtained possession of a lucifer match, and struck a light by striking the match against the wall. Instantly there was a blaze. Fortunately for him, in his fright, he threw the match on the floor. His mother at this moment entered the room. If his clothes had taken fire, which they might have done, had he not have thrown the match away, or if his mother had not been so near at hand, he would, in all probability, have either been severely burned or have been burned to death.

290. *If a child's clothes take fire, what ought to be done to extinguished them ?*

Lay him on the floor, then roll him either in the rug, or in the carpet, or in the door-mat, or in any thick article of dress you may either have on, or have at hand —if it be woollen, so much the better ; or, throw him down, and roll him over and over on the floor, as, by excluding the atmospheric air, the flame will go out :— hence the importance of a mother cultivating presence of mind. If parents were better prepared for such emergencies, such horrid disfigurations and frightful deaths would be less frequent.

You ought to have a proper fire-guard before the nursery grate, and should be strict in not allowing your child to play with fire. If he still persevere in playing with it, when he has been repeatedly cautioned not to do so, he should be punished for his temerity. If anything would justify corporal chastisement, it would surely be such an act of disobedience. There are only two acts of disobedience that I would flog a child for—namely, the playing with fire and the telling of a lie ! If after various warnings and wholesome corrections he still persist, it would be well to let him slightly taste the pain of his doing so, either by holding his hand for a moment very near the fire, or by allowing him to slightly touch either the hot bar of the grate or the flame of the candle. Take my word for it the above plan will effectually cure him—he will never do it again. It would be well for the children of the poor to have pinafores made either of woollen or of stuff materials. The dreadful deaths from burning, which so often occur in winter, too frequently arise from *cotton* pinafores first taking fire.*

If all dresses after being washed, and just before being dried, were, for a short time, soaked in a solution of tungstate of soda, such clothes, when dried, would be perfectly fire-proof.

Tungstate of soda may be used either with or without starch; but full directions for the using of it will, at the time of purchase, be given by the chemist.

291. *Is a burn more dangerous than a scald ?*

A burn is generally more serious than a scald. Burns and scalds are more dangerous on the body, especially on the chest, than either on the face or on the extremities. The younger the child, the greater the danger.

Scalds both of the mouth and the throat, from a child drinking boiling water from the spout of a tea-kettle, are most dangerous. A poor person's child is, from the unavoidable absence of the mother, sometimes shut up in the kitchen by himself, and being very thirsty, and no other water being at hand, he is tempted, in his ignorance, to drink from the tea-kettle : If the water be unfortunately boiling, it will most likely prove to him to be a fatal draught !

292. *What are the best immediate applications to a scald or to a burn ?*

There is nothing more efficacious than flour. It ought to be thickly applied over the part affected, and should be kept in its place either with a rag and a bandage, or with strips of old linen. If this be done, almost instantaneous relief will be experienced, and the burn or the scald, if superficial, will soon be well. The advantage of flour as a remedy, is this, that it is always at hand. I have seen some extensive burns and scalds cured by the above simple plan. Another excellent remedy is, cotton-wool of superior quality, purposely made for surgeons. The burn or the scald ought to be enveloped in it; layer

---

* "It has been computed that upwards of 1000 children are annually burned to death by accident in England.

after layer should be applied until it be several inches thick. The cotton-wool must not be removed for several days. These two remedies, flour and cotton-wool, may be used in conjunction; that is to say, the flour may be thickly applied to the scald or to the burn, and the cotton wool over all.

Prepared lard—that is to say, lard without salt*—is an admirable remedy for burns and for scalds. The advantages of lard are,—(1.) It is almost always at hand; (2.) It is very cooling, soothing, and unirritating to the part, and it gives almost immediate freedom from pain; (3.) It effectually protects and sheathes the burn or the scald from the air; (4.) It is readily and easily. applied: all that has to be done is to spread the lard either on pieces of old linen rag, or on lint, and then to apply them smoothly to the parts affected, keeping them in their places by means of bandages—which bandages may be readily made from either old linen or calico shirts. Dr John Packard, of Philadelphia, was the first to bring this remedy for burns and scalds before the public—he having tried it in numerous instances, and with the happiest results. I myself have, for many years been in the habit of prescribing lard as a dressing for blisters, and with the best effects. I generally advise equal parts of prepared lard and of spermaceti-cerate to be blended together to make an ointment. The spermaceti-cerate gives a little more consistence to the lard, which, in warm weather especially, is a great advantage.

Another valuable remedy for burns is "carron-oil;" which is made by mixing equal parts of linseed-oil and lime-water in a bottle, and shaking it up before using it.

Cold applications, such as cold water, cold vinegar and

---

* If there be no other lard in the house but lard *with* salt, the salt may be readily removed by washing the lard in cold water. Prepared lard—that is to say, lard *without* salt—can, at any moment, be procured from the nearest druggist in the neighbourhood

water, and cold lotions, are most injurious, and, in many cases, even dangerous. Scraped potatoes, sliced cucumber, salt, and spirits of turpentine, have all been recommended ; but, in my practice, nothing has been so efficacious as the remedies above enumerated.

Do not wash the wound, and do not dress it more frequently than every *other* day. If there be much discharge, let it be gently sopped up with soft old linen rag ; but do not, *on any* account, let the burn be rubbed or roughly handled. I am convinced that, in the majority of cases, wounds are too frequently dressed, and that the washing of wounds prevents the healing of them. "It is a great mistake," said Ambrose Paré, "to dress ulcers too often, and to wipe their surfaces clean, for thereby we not only remove the useless excrement, which is the mud or sanies of ulcers, but also the matter which forms the flesh. Consequently, for these reasons, ulcers should not be dressed too often."

It is nature, and not the surgeon, that really cures the wound, and it is done, like all Nature's works, principally in secret, by degrees, and by patience, and resents much interference. The seldom-dressing of a wound and patience are, then, two of the best remedies for effecting a cure. Shakspeare, who seemed to know surgery, as he did almost everything else besides, was quite cognisant of the fact :—

"How poor are they, that have not patience !
What wound did ever heal, but by degrees ?"

The burn or the scald may, after the first two days, if severe, require different dressings ; but, if it be severe, the child ought of course to be immediately placed under the care of a surgeon.

If the scald be either on the leg or on the foot, a common practice is to take the shoe and the stocking off ; in this operation the skin is also at the same time very apt to be removed. Now, both the shoe and the stocking ought to be slit up, and thus be taken off, so that neither unnecessary pain nor mischief may be caused.

Q

293. *If a bit of quick-lime should accidentally enter the eye of my child, what ought to be done?*

Instantly, but tenderly remove, either by means of a camel's hair brush, or by a small spill of paper, any bit of lime that may adhere to the ball of the eye, or that may be within the eye or on the eye-lashes; then well bathe the eye (allowing a portion to enter it) with vinegar and water—one part of vinegar to three parts of water, that is to say, a quarter fill a clean half-pint medicine bottle with vinegar, and then fill it up with spring water, and it will be ready for use. Let the eye be bathed for at least a quarter of an hour with it. The vinegar will neutralise the lime, and will rob it of its burning properties.

Having bathed the eye with vinegar and water for a quarter of an hour, bathe it for another quarter of an hour simply with a little warm water, after which, drop into the eye two or three drops of the best sweet-oil, put on an eye-shade made of three thicknesses of linen rag, covered with green silk, and then do nothing more until the doctor arrive.

If the above rules be not *promptly* and *properly* followed out, the child may irreparably lose his eyesight; hence the necessity of conversations of this kind, to tell a mother, provided *immediate* assistance cannot be obtained, what ought *instantly* to be done; for moments, in such a case, are precious.

While doing all that I have just recommended, let a surgeon be sent for, as a smart attack of inflammation of the eye is very apt to follow the burn of lime; but which inflammation will, provided the *previous* directions have been *promptly* and *efficiently* followed out, with appropriate treatment, soon subside.

The above accident is apt to occur to a child who is standing near a building when the slacking of quick-lime is going on, and where portions of lime in the form of powder are flying about the air. It would be well not to allow a child to stand about such places, as prevention is always better than cure. *Quick lime* is sometimes called *caustic-lime :* it well deserves its name, for it is a *burning-lime,* and if proper means be not promptly used, will soon burn away the sight.

294. *If any other foreign substance should enter the eye, what is the best method of removing it ?*

If there be grit, or sand, or dust, or particle of coal, or gnat, or a hair, or an eye-lash in the eye, it ought to be tenderly removed by a small tightly-folded paper spill, holding down the lower lid with the fore-finger of the left hand the while ; and the eye, if inflamed, should be frequently bathed with warm milk and water ; but generally as soon as the cause is removed the effect will cease, and after treatment will be unnecessary.

If a particle of metal be sticking on the cornea of the eye, as it sometimes does, it will require the skilled hand of a surgeon to remove it.

Any foreign substance, however minute, in the eye, is very painful ; but a piece of burning lime is excruciating. Shakspeare gives a graphic description of the pain from the presence of any foreign substance, however small, in the eye :—

> "O heaven !—that there were but a mote in yours,
> A grain, a dust, a gnat, a wand'ring hair,
> Any annoyance in that precious sense !
> Then, feeling what small things are boist'rous there,
> Your vile intent must needs seem horrible."

295. *What ought to be done in a case of choking ?*

How often does a hungry little child, if not carefully watched, fill his mouth so full, and swallow lumps of food in such hot haste, as to choke himself—

> "With eager feeding, food doth choke the feeder."
> *Shakspeare.*

Treatment.—Instantly put your finger into the throat and feel if the substance be within reach ; if it be food, force it down, and thus liberate the breathing ; should it be a hard substance, endeavour to hook it out ; if you cannot reach it, give a good smart blow or two with the flat of the hand on the back ; or, as recommended by a

contributor to the *Lancet*, on the chest, taking care to
" seize the little patient, and place him between your
knees side ways, and in this or some other manner to
compress the abdomen [the belly], otherwise the power
of the blow will be lost by the yielding of the abdominal
parieties [walls of the belly], and the respiratory effort will
not be produced." If that does not have the desired
effect, tickle the throat with your finger, so as to ensure
immediate vomiting, and the subsequent ejection of the
offending substance.

296. *Should my child be bitten by a dog supposed to
be mad, what ought to be done ?*

Instantly well rub for the space of five or ten *seconds*—
seconds, *not* minutes—a stick of nitrate of silver (lunar-
caustic) into the wound. The stick of lunar-caustic
should be pointed, like a cedar pencil for writing, in
order the more thoroughly to enter the wound.* This,
if properly done directly after the bite, will effectually
prevent hydrophobia. The nitrate of silver acts not
only as a caustic to the part, but it appears effectually
to neutralise the poison, and thus, by making the virus
perfectly innocuous, is a complete antidote. If it be
either the lip, or the parts near the eye, or the wrist,
that have been bitten, it is far preferable to apply the
caustic than to cut the part out ; as the former is neither
so formidable, nor so dangerous, nor so disfiguring as
the latter, and yet it is equally as efficacious. I am indebted to the late Mr Youatt, the celebrated veterinary
surgeon, for this valuable antidote or remedy for the
prevention of the most horrible, heart-rending, and incurable disease known. Mr Youatt had an immense
practice among dogs as well as among horses. He was
a keen observer of disease, and a dear lover of his profession, and he had paid great attention to rabies—dog-
madness. He and his assistants had been repeatedly
bitten by rabid dogs ; but knowing that he was in pos-

---

* A stick of *pointed* nitrate of silver, in a case, ready for use,
may be procured of any respectable chemist.

session of an infallible preventive remedy, he never dreaded the wounds inflicted either upon himself or upon his assistants. Mr Youatt never knew lunar-caustic, if properly and *immediately* applied, to fail. It is, of course, only a preventive. If hydrophobia be once developed in the human system, no antidote has ever yet, for this fell and intractable disease, been found.

While walking the London Hospitals, upwards of forty years ago, I received an invitation from Mr Youatt to attend a lecture on rabies—dog-madness. He had, during the lecture, a dog present labouring under *incipient* madness. In a day or two after the lecture, he requested me and other students to call at his infirmary and see the dog, as the disease was at that time fully developed. We did so, and found the poor animal raving mad—frothing at the mouth, and snapping at the iron bars of his prison. I was particularly struck with a peculiar brilliancy and wildness of the dog's eyes. He seemed as though, with affright and consternation, he beheld objects unseen by all around. It was pitiful to witness his frightened and anxious countenance. Death soon closed the scene !

I have thought it my duty to bring the value of lunar-caustic as a preventive of hydrophobia prominently before your notice, and to pay a tribute of respect to the memory of Mr Youatt—a man of talent and of genius.

Never kill a dog supposed to be mad who has bitten either a child, or any one else, until it has, past all doubt, been ascertained whether he be really mad or not. He ought, of course, to be tied up ; and be carefully watched, and be prevented the while from biting any one else. The dog by all means should be allowed to live at least for some weeks, as the fact of his remaining well will be the best guarantee that there is no fear of the bitten child having caught hydrophobia.

There is a foolish prejudice abroad, that a dog, be he mad or not, who has bitten a person ought to be *immediately* destroyed ; that although the dog be not at the time mad, but should at a future period become so, the person who had been bitten when the dog was *not* mad, would, when the dog became mad, have hydrophobia ! It seems almost absurd to bring the subject forward ; but the opinion is so very general and deep-rooted, that I think it well to declare that there is not the slightest foundation of truth in it, but that it is a ridiculous fallacy !

A cat sometimes goes mad, and its bite may cause hydrophobia ; indeed, the bite of a mad cat is more dangerous than the bite of a mad dog. A bite from a mad cat ought to be treated precisely in the same manner—namely, with the lunar-caustic—as for a mad dog.

Hydrophobia was by our forefathers graphically called *water-fright :* it was well named, for the horror of swallowing water is, by an hydrophobic patient, most intense, and is *the* leading symptom of this fell and incurable disease.

A bite either from a dog or from a cat *who is not mad*, from a cat especially, is often venomous and difficult to heal. The best application is, *immediately* to apply a large hot white bread poultice to the part, and to renew it every four hours ; and, if there be much pain in the wound, to well foment the part, every time before applying the poultice, with a hot camomile and poppy-head fomentation.

Scratches of a cat are best treated by smearing, and that freely and continuously for an hour, and then afterwards at longer intervals, fresh butter on the part affected. If fresh butter be not at hand, fresh lard—that is to say, lard *without* salt—will answer the purpose. If the pain of the scratch be very intense, foment the part affected with hot water, and then apply a hot white bread poultice, which should be frequently renewed.

297. *What are the best remedies in case of a sting from either a bee or a wasp ?*

Extract the sting, if it have been left behind, either by means of the pair of dressing forceps, or by the pressure of the hollow of a small key—a watch-key will answer the purpose ; then, the blue-bag (which is used in washing) moistened with water, should be applied to the part ; or a few drops of solution of potash,* or " apply moist snuff or tobacco, rubbing it well in,"† and renew from time to time either of them : if either of these be not at hand, either honey, or treacle, or fresh butter, will answer· the purpose. Should there be much swelling or inflammation, foment the part with hot water, and then apply hot bread poultice, and renew it frequently. In eating apricots, or peaches, or other fruit, they ought beforehand to be carefully examined, in order to ascertain that no wasp is lurking in them ; otherwise, it may sting the throat, and serious consequences will ensue.

298. *If a child receive a fall, causing the skin to be grazed, can you tell me of a good application ?*

You will find gummed paper an excellent remedy : the way of preparing it is as follows :—Apply evenly, by means of a small brush, thick mucilage of gum-arabic to cap-paper ; hang it up to dry, and keep it ready for use. When wanted, cut a portion as large as may be requisite, then moisten it with your tongue, in the same manner you would a postage stamp, and apply it to the grazed part. It may be removed when necessary by simply wetting it· with water. The part in two or three days will be well. There is usually a margin of gummed paper sold with postage stamps ; this will answer the purpose equally well. If the gummed paper be not at hand, then frequently, for the space of an hour or two, smear the part affected with fresh butter.

299. *In case of a child swallowing by mistake either laudanum, or paregoric, or Godfrey's Cordial, or any other preparation of opium, what ought to be done ?*

Give, as *quickly as possible,* a strong mustard emetic ; that is to say, mix two tea-spoonfuls of flour of mustard in half a tea-cupful of water, and force it down his throat. If free vomiting be not induced, tickle the

---

* Which may be instantly procured of a druggist.
† A Bee-master. *The Times,* July 28, 1864.

upper part of the swallow with a feather; drench the little patient's stomach with large quantities of warm water. As soon as it can be obtained from a druggist, give him the following emetic draught :—

> Take of—Sulphate of Zinc, one scruple;
> Simple Syrup, one drachm ;
> Distilled Water, seven drachms ;
> To make a Draught.

Smack his buttocks and his back; walk him, or lead him, or carry him about in the fresh air; shake him by the shoulders; pull his hair; tickle his nostrils; shout and holla in his ears ; plunge him into a warm bath and then into a cold bath alternately; well sponge his head and face with cold water ; dash cold water on his head, face, and neck ; and do not, on any account, until the effects of the opiate are gone off, allow him to go to sleep ; if you do, he will never wake again ! While doing all these things, of course, you ought to lose no time in sending for a medical man.

300. *Have you any observation to make on parents allowing the Deadly Nightshade—the Atropa Belladonna—to grow in their gardens ?*

I wish to caution you *not* on any account to allow the Belladonna—the Deadly Nightshade—to grow in your garden. The whole plant—root, leaves, and berries—is poisonous ; and the berries, being attractive to the eye, are very alluring to children.

301. *What is the treatment of poisoning by Belladonna?*

Instantly send for a medical man ; but, in the meantime, give an emetic—a mustard emetic :—mix two teaspoonfuls of flour of mustard in half a tea-cupful of warm water, and force it down the child's throat : then drench him with warm water, and tickle the upper part of his swallow either with a feather or with the finger, to make him sick : as the grand remedy is an emetic to bring up the offending cause. If the emetic have not acted sufficiently, the medical man when he arrives may deem it necessary to use the stomach-pump ; but remember not a moment must be lost, for moments are precious in a case of belladonna poisoning, in giving a mustard emetic, and repeating it again and again until the enemy be dislodged. Dash cold water upon his head and face : the best way of doing which is by means of a large sponge, holding his head and his face over a wash-hand basin, half filled with cold water, and filling the sponge from the basin, and squeezing it over his head and face, allowing the water to continuously stream over them for an hour or two, or until the effects of the poison have passed away. This sponging of the head and face is very useful in poisoning by opium, as well as in poisoning by belladonna ; indeed, the treatment of poisoning by the one is very similar to the treatment of poisoning by the other. I, therefore, for the further treatment of poisoning by belladonna, beg to refer you to a previous Conversation on the treatment of poisoning by opium.

302. *Should a child put either a pea or a bead, or any other foreign substance, up the nose, what ought to be done ?*

Do not attempt to extract it yourself, or you might push it further in, but send instantly for a surgeon, who will readily remove it, either with a pair of forceps, or by means of a bent probe, or with a director. If it be a pea, and it be allowed for any length of time to remain in, it will swell, and will thus become difficult to extract, and may produce great irritation and inflammation. A child ought not to be allowed to play with peas or with beads (unless the beads are on a string), as he is apt, for amusement, to push them up his nose.

303. *If a child have put either a pea, a bean, a bead, a cherry-stone, or any other smooth substance, into his ear, what ought to be done to remove it ?*

Turn his head on one side, in order to let the ear with the pea or the bead in it be undermost, then give with the flat of your hand two or three sharp, sudden slaps or boxes on the other, or *uppermost* ear, and most likely the offending substance will drop out. Poking at the ear will, in the majority of cases, only send the substance further in, and will make it more difficult (if the above simple plan does not succeed) for the medical man to

remove. The surgeon will, in all probability, syringe
the ear ; therefore have a supply of warm water in readiness for him, in order that no time may be lost.

304. *If an earwig or any other living thing, should get
into the ear of a child, what ought to be done ?*

Lay the child on his side, the affected ear being uppermost, and fill the ear, from a tea-spoon, with either
water or sweet oil. The water or oil will carry the living thing, whatever it be, out of the ear, and the child
is at once relieved.

305. *If a child swallow a piece of broken glass, what
ought to be done ?*

Avoid purgatives, as the free action on the bowels
would be likely to force the spiculæ of glass into the
mucous membrane of the bowels, and thus would wound
them, and might cause ulceration, and even death.
"The object of treatment will be to allow them to pass
through the intestines well enveloped by the other contents of the tube ; and for this purpose a solid,
farinaceous diet should be ordered, and purgatives
scrupulously avoided."—*Shaw's Medical Remembrancer,*
by Hutchinson.

306. *If a child swallow a pin, what should be done ?*

Treat him as for broken glass. Give him no aperients,
or it might, in action, force the pin into the bowel. I
have known more than one instance where a child, after
swallowing a pin, to have voided it in his motion.

307. *If a child swallow a coin of any kind, is danger
likely to ensue, and what ought to be done ?*

There is, as a rule, no danger. A dose or two of castor
oil will be all that is usually necessary. The evacuations
ought to be carefully examined until the coin be discovered. I once knew a child swallow a pennypiece,
and pass it in his stool.

308. *If a child, while playing with a small coin (such
as either a threepenny or a fourpenny piece), or any
other substance, should toss it into his mouth, and inad-
vertently allow it to enter the windpipe, what ought to be
done ?*

Take hold of him by the legs, allowing his head to hang downwards ; then give him with the palm of your hand several sharp blows on his back, and you may have the good fortune to see the coin coughed out of his mouth. Of course, if this plan does not succeed, send instantly for a medical man.

309. *How can a mother prevent her child from having an accident ?*

By strict supervision over him on her own part, and by not permitting her child to be left to the tender mercies of servants ; by not allowing him to play with fire, to swing over banisters, and to have knives and playthings of a dangerous character ; to keep all poisonous articles and cutting instruments out of his reach ; and, above all and before all, insisting, lovingly, affectionately, but firmly, upon implicit obedience.

Accidents generally arise from one of three causes, namely, either from wilful disobedience, or from gross carelessness, or from downright folly. I quite agree with Davenant, that they do not arise from chance—

> "If we consider accident,
>   And how, repugnant unto sense,
> It pays desert with bad event,
>   We shall disparage Providence."

PART III.

BOYHOOD AND GIRLHOOD.

Just at the age 'twixt boy and youth.
When thought is speech, and speech is truth.—SCOTT.

'Tis. with him e'en standing water.
Between man and boy. — SHAKSPEARE.

Standing with reluctant feet.
Where the brook and river meet.
Womanhood and childhood fleet!—LONGFELLOW.

ABLUTION, ETC.

310. *Have you any remarks to make on the ablution of boys and girls?*

How is it that a mother thinks it absolutely necessary (which it really is) that her babe's *whole* body should, every morning, be washed ; and yet who does not deem it needful that her girl or boy, of twelve years old, should go through the process of daily and *thorough* ablution? If the one case be necessary, sure I am that the other is equally if not more needful.

Thorough ablution of the body every morning at least is essential to health. I maintain that no one can be in the enjoyment of perfect health who does not keep his skin—the whole of his skin—clean. In the absence of cleanliness, a pellicle forms on the skin which engenders disease. Moreover, a person who does not keep his skin clean is more susceptible of contracting contagious disease, such as small-pox, typhus fever, cholera, diphtheria, scarlet fever, &c.

Thorough ablution of the body is a grand requisite of

health. I maintain that no one can be perfectly healthy unless he thoroughly wash his body—the whole of his body ; if filth accumulate which, if not washed off, it is sure to do, disease must, as a matter of course, follow. Besides, ablution is a delightful process ; it makes one feel fresh and sweet, and young and healthy ; it makes the young look handsome, and the old look young ! Thorough ablution might truly be said both to renovate and to rejuvenise ! A scrupulously clean skin is one of the grand distinctive characteristics both of a lady and of a gentleman.

Dirty people are not only a nuisance to themselves, but to all around ; they are not only a nuisance but a danger, as their dirty bodies are apt to carry from place to place contagious diseases.

It is important that parts that are covered should be kept cleaner than parts exposed to the air, as dirt is more apt to fester in dark places ; besides, parts exposed to the air have the advantage of the air's sweetening properties ; air acts as a bath, and purifies the skin amazingly.

It is desirable to commence a complete system of washing early in life, as it then becomes a second nature, and cannot afterwards be dispensed with. One accustomed to the luxury of his morning ablution, if anything prevented him from taking it, would feel most uncomfortable ; he would as soon think of dispensing with his breakfast as with his bath.

Every boy, every girl, and every adult, ought each to have either a room or a dressing-room to himself or to herself, in order that he or she might strip to the skin and thoroughly wash themselves ; no one can wash properly and effectually without doing so.

Now, for the paraphernalia required for the process :— (1.) A large nursery basin, one that will hold six or eight quarts of water (Wedgwood's make being considered the best) ; (2.) A piece of coarse flannel, a yard long and half a yard wide ; (3.) A large sponge ; (4.) A tablet either of the best yellow or of curd soap ; (5.) Two towels—one being a diaper, and the other a Turkish rubber. Now, as to the manner of performing ablution. You ought to fill the basin three parts full with *rain* water; then, having well-soaped and cleansed your hands, re-soap them, dip your head and face into the water, then with the soaped hands well rub and wash your head, face, neck, chest, and armpits; having done which, take the wetted sponge, and go over all the parts previously travelled over by the soaped hands; then fold the flannel as you would a neck-kerchief, and dip it in the water, then throw it, as you would a skipping-rope, over your shoulders and move it a few times from right to left and from left to right, and up and down, and then across the back and loins; having done which, dip the sponge in the water, and holding your head over the water, let the water stream from the sponge a time or two over your head, neck, and face. Dip your head and face in the water, then put your hands and arms (as far as they will go) into the water, holding them there while you can count thirty. Having reduced the quantity of water to a third of a basinful, place the basin on the floor, and sit (while you can count fifty) *in* the water; then put one foot at a time in the water, and quickly rub, with soaped hands, up and down your leg, over the foot, and pass your thumb between each toe (this latter procedure tends to keep away soft corns); then take the sponge, filled with water, and squeeze it over your leg and foot, from the knee downwards,—then serve your other leg and foot in the same way. By adopting the above plan, the whole of the body will, every morning, be thoroughly washed.

A little warm water might at first, and during the winter time, be added, to take off the chill; but the sooner quite cold water is used the better. The body ought to be quickly dried (taking care to wipe between each toe), first with the diaper, and then with the Turkish rubber. In drying your back and loins, you ought to throw as you would a skipping-rope, the Turkish rubber over your shoulders, and move it a few times from side to side, until the parts be dry.

Although the above description is necessarily prolix, the washing itself ought to be very expeditiously performed; there should be no dawdling over it, otherwise the body will become chilled, and harm instead of good will be the result. If due dispatch be used, the whole of the body might, according to the above method, be thoroughly washed and dried in the space of ten minutes.

A boy ought to wash his head, as above directed, every morning, a girl, who has much hair, once a week, with soap and water, with flannel and sponge. The hair, if not frequently washed, is very dirty, and nothing is more repulsive than a dirty head !

It might be said, " Why do you go into particulars ? why dwell so much upon minutiæ ? Every one, without being told, knows how to wash himself ! " I reply, " That very few people do know how to wash themselves properly ; it is a misfortune that they do not—they would be healthier and happier and sweeter if they did !"

611. *Have you any remarks to make on boys and girls learning to swim ?*

Let me strongly urge you to let your sons and daughters be *early* taught to swim. Swimming is a glorious exercise—one of the best that can be taken; it expands the chest; it promotes digestion; it develops the muscles, and brings into action some muscles that in any other form of exercise are but seldom brought into play; it strengthens and braces the whole frame, and thus makes the swimmer resist the liability of catching cold; it gives both boys and girls courage, energy, and self-reliance,—splendid qualities in this rough world of ours. Swimming is oftentimes the means of saving human life; this of itself would be a great recommendation of its value. It is a delightful amusement; to breast the waves is as exhilarating to the spirits as clearing on horse-back a five-barred gate.

The art of learning to swim is quite as necessary to be learned by a girl as by a boy ; the former has similar muscles, lungs, and other organs to develop as the latter,

It is very desirable that in large towns swimming-baths for ladies should be instituted. Swimming ought, then, to be a part and parcel of the education of every boy and of every girl.

Swimming does not always agree. This sometimes arises from a person being quite cold before he plunges into the water. Many people have an idea that they ought to go into the water while their bodies are in a cool state. Now this is a mistaken notion, and is likely to produce dangerous consequences. The skin ought to be comfortably warm, neither very hot nor very cold, and then the bather will receive every advantage that cold bathing can produce. If he go into the bath whilst the body is cold, the blood becomes chilled, and is driven to internal parts, and thus mischief is frequently produced.

A boy, after using cold bathing, ought, if it *agree* with him, to experience a pleasing glow over the whole surface of his body, his spirits and appetite should be increased, and he ought to feel stronger; but if it *disagree* with him, a chilliness and coldness, a lassitude and a depression of spirits, will be the result ; the face will be pale and the features will be pinched, and, in some instances, the lips and the nails will become blue ; all these are signs that *cold* bathing is injurious, and, therefore, that it ought on no account to be persevered in, unless these symptoms have hitherto proceeded from his going into the bath whilst he was quite cold. He may, previously to entering the bath, warm himself by walking briskly for a few minutes. Where *cold* sea water bathing does not agree, *warm* sea bathing should be substituted.

312. *Which do you prefer—sea bathing or fresh water bathing ?*

Sea bathing. Sea bathing is incomparably superior to fresh water bathing ; the salt water is far more refreshing and invigorating ; the battling with the waves is more exciting ; the sea breezes, blowing on the nude body, breathes (for the skin is a breathing apparatus) health and strength into the frame, and comeliness into the face ; the sea water and the sea breezes are splendid cosmetics ; the salt water is one of the finest applications, both for strengthening the roots and brightening the colour of the hair, provided grease and pomatum have not been previously used.

313. *Have you any directions to give as to the time and the seasons, and the best mode of sea bathing ?*

Summer and autumn are the best seasons of the year for cold sea bathing—August and September being the best months. To prepare the skin for the cold sea bathing, it would be well, before taking a dip in the sea, to have on the previous day a warm salt water bath. It is injurious, and even dangerous, to bathe *immediately* after a *full* meal ; the best time to bathe is about two hours after breakfast—that is to say, at about eleven or twelve o'clock in the forenoon. The bather as soon as he enters the water, ought *instantly* to wet his head ; this may be done either by his jumping at once from the machine into the water, or, if he have not the courage to do so, by plunging his head without loss of time *completely* under the water. He should remain in the water about a quarter of an hour, but never longer. than half an hour. Many bathers by remaining a long time in the water do themselves great injury. If sea bathing be found to be invigorating—and how often to the delicate it has proved to be truly magical—a patient may bathe once every day, but on no account oftener. If he be not strong, he had better, at first, bathe only every other day, or even only twice a week. The bather, after leaving the machine, ought for half an hour to take a brisk walk in order to promote a reaction, and thus to cause a free circulation of the blood.

314. *Do you think a tepid bath* may be more safely used?*

A tepid bath may be taken at almost any time, and a bather may remain longer in one, with safety, than in a cold bath.

---

* A tepid bath from 62 to 96 degrees of Fahrenheit.

R

315. *Do you approve of warm bathing?*

A warm bath* may with advantage be occasionally used—say, once a week. A warm bath cleanses the skin more effectually than either a cold or a tepid bath; but, as it is more relaxing, ought not to be employed so often as either of them. A person should not continue longer than ten minutes in a warm bath. Once a week, as a rule, is quite often enough for a warm bath; and it would be an excellent plan if every boy and girl and adult would make a practice of having one *regularly* every week, unless any special reason should arise to forbid its use.

316. *But does not warm bathing, by relaxing the pores of the skin, cause a person to catch cold if he expose himself to the air immediately afterwards?*

There is, on this point, a great deal of misconception and unnecessary fear. A person, *immediately* after using a warm bath, should take proper precautions—that is to say, he must not expose himself to draughts, neither ought he to wash himself in *cold* water, nor should he, *immediately* after taking one, drink *cold* water. But he may follow his usual exercise or employment, provided the weather be fine, and the wind be neither in the east nor the north-east.

Every house of any pretension ought to have a bath-room. Nothing would be more conducive to health than regular systematic bathing. A hot and cold bath, a sitz bath, and a shower bath—each and all in their turn—are grand requisites to preserve and procure health. If the house cannot boast of a bath-room, then the Corporation Baths (which nearly every large town possesses) ought to be liberally patronised.

MANAGEMENT OF THE HAIR.

317. *What is the best application for the hair?*

A sponge and *cold* water, and two good hair-brushes. Avoid grease, pomatum, bandaline, and all abominations

---

* A warm bath from 97 to 100 degrees of Fahrenheit

of that kind. There is a natural oil of the hair, which is far superior to either Rowland's Macassar Oil or any other oil! The best scent for the hair is an occasional dressing of soap and water; the best beautifier of the hair is a downright thorough good brushing with two good hair brushes! Again, I say, *avoid grease of all kinds to the hair.* " And as for woman's hair, don't plaster it with scented and sour grease, or with any grease; it has an oil of its own. And don't tie up your hair tight, and make it like a cap of iron over your skull. And why are your ears covered? You hear all the worse, and they are not the cleaner. Besides, the ear is beautiful in itself, and plays its own part in the concert of the features."*

If the hair cannot, without some application, be kept tidy, then a little castor oil, scented, might, by means of an old tooth-brush, be used to smooth it; castor oil is, for the purpose, one of the most simple and harmless of dressings; but, as I said before, the hair's own natural oil cannot be equalled, far less surpassed!

If the hair fall off, the castor oil, scented with a few drops either of otto of roses or of essence of bergamot, is a good remedy to prevent its doing so; a little of it ought, night and morning, to be well rubbed into the roots of the hair. Cocoa-nut oil is another excellent application for the falling off of the hair, and can never do harm, which is more than can be said of many vaunted remedies for the hair!

CLOTHING.

318. *Do you approve of a boy wearing flannel next the skin?*

England is so variable a climate, and the changes from heat to cold, and from dryness to moisture of the atmosphere, are so sudden, that some means are required to guard against their effects. Flannel, as it is a bad conductor of heat, prevents the sudden changes from affecting the body, and thus is a great preservative against cold.

---

* *Health.* By John Brown, M.D.

Flannel is as necessary in the summer as in the winter time ; indeed, we are more likely both to sit and to stand in draughts in the summer than in the winter ; and thus we are more liable to become chilled and to catch cold.

Woollen shirts are now much worn ; they are very comfortable and beneficial to health. Moreover, they simplify the dress, as they supersede the necessity of wearing either both flannel and linen, or flannel and calico shirts.

319. *Flannel sometimes produces great irritation of the skin : what ought to be done to prevent it ?*

Have a moderately fine flannel, and persevere in its use ; the skin in a few days will bear it comfortably. The Angola and wove-silk waistcoats have been recommended as substitutes, but there is nothing equal to the old-fashioned Welsh flannel.

320. *If a boy have delicate lungs, do you approve of his wearing a prepared hare-skin over the chest ?*

I do not : the chest may be kept too warm as well as too cold. The hare-skin heats the chest too much, and thereby promotes a violent perspiration ; which, by his going into the cold air, may become suddenly checked, and may thus produce mischief. If the chest be delicate, there is nothing like flannel to ward off colds.

321. *After an attack of Rheumatic Fever, what extra clothing do you advise ?*

In the case of a boy, or a girl, just recovering from a severe attack of Rheumatic Fever, flannel next the skin ought always, winter and summer, to be worn—flannel drawers as well as a flannel vest.

322. *Have you any remarks to make on boys' waist-coats ?*

Fashion in this, as in most other instances, is at direct variance with common sense. It would seem that fashion was intended to make work for the doctor, and to swell the bills of mortality ! It might be asked, What part of the chest, in particular, ought to be kept warm ? The upper part needs it most. It is in the *upper* part of the lungs that tubercles (consumption) usually first make their appearance; and is it not preposterous to have such parts, in particular, kept cool? Double-breasted waistcoats cannot be too strongly recommended for *delicate* youths, and for all men who have *weak* chests.

323. *Have you any directions to give respecting the shoes and the stockings?*

The shoes for winter should be moderately thick and waterproof. If boys and girls be delicate, they ought to have double soles to their shoes, with a piece of bladder between each sole, or the inner sole may be made of cork; either of the above plans will make the soles of boots and shoes completely water-proof. In wet or dirty weather India-rubber over-shoes are useful, as they keep the *upper* as well as the *under* leathers perfectly dry.

The socks, or stockings, for winter, ought to be either lambs-wool or worsted; it is absurd to wear *cotton* socks or stockings all the year round. I should advise a boy to wear socks not stockings, as he will then be able to dispense with garters. Garters, as I have remarked in a previous Conversation, are injurious—they not only interfere with the circulation of the blood, but also, by pressure, injure the bones, and thus the shape of the legs.

Boys and girls cannot be too particular in keeping their feet warm and dry, as cold wet feet are one of the most frequent exciting causes of bronchitis, of sore throats, and of consumption.

324. *When should a girl begin to wear stays?*

She ought never to wear them.

325. *Do not stays strengthen the body?*

No; on the contrary, they weaken it. (1.) *They weaken the muscles.* The pressure upon them causes them to waste; so that, in the end, a girl cannot do without them, as the stays are then obliged to perform the duty of the wasted muscles. (2.) *They weaken the lungs* by interfering with their functions. Every inspiration is accompanied by a movement of the ribs. If this movement be impeded, the functions of the lungs are impeded likewise; and, consequently, disease is likely to follow ; and either difficulty of breathing, or cough, or consumption, may ensue. (3.) *They weaken the heart's action*, and thus frequently produce palpitation, and, perhaps, eventually, organic or incurable disease of the heart. (4.) *They weaken the digestion*, by pushing down the stomach and the liver, and by compressing the latter ; and thus induce indigestion, flatulence, and liver-disease.* (5.) *They weaken the bowels*, by impeding their proper peristaltic (spiral) motion, and thus might produce either constipation or a rupture. Is it not presumptuous to imagine that man can improve upon God's works ; and that if more support had been required, the Almighty would not have given it ?—

" God never made his work for man to mend."—*Dryden.*

326. *Have you any remarks to make on female dress ?*
There is a perfect disregard of health in everything appertaining to fashion. Parts that ought to be kept warm, remain unclothed ; the *upper* portion of the chest, most prone to tubercles (consumption), is completely exposed ; the feet, great inlets to cold, are covered with thin stockings, and with shoes as thin as paper. Parts that should have full play are cramped and hampered ; the chest is cribbed in with stays, the feet with *tight* shoes,—hence causing deformity, and preventing a free circulation of blood. The mind, that ought to be calm and unruffled, is kept in a constant state of excitement by balls, and concerts, and plays. Mind and body sympathise with each other, and disease is the consequence. Night is turned into day ; and a delicate girl leaves the heated ball-room, decked out in her airy finery, to breathe the damp and cold air of night. She goes to bed, but, for the first few hours, she is too much excited to sleep ; towards morning, when the air is pure and invigorating, and, when to breathe it, would be to

---

* Several years ago, while prosecuting my anatomical studies in London University College Dissecting-rooms, on opening a young woman, I discovered an immense indentation of the liver large enough to admit a rolling-pin, produced by tight-lacing !

inhale health and life, she falls into a feverish slumber, and wakes not until noon-day. Oh, that a mother should be so blinded and so infatuated !

327. *Have you any observations to make on a girl wearing a green dress ?*

It is injurious to wear a *green* dress, if the colour have been imparted to it by means of *Scheele's green*, which is arsenite of copper—a deadly poison. I have known the arsenic to fly off from a *green* dress in the form of powder, and to produce, in consequence, ill-health. Gas-light green is a lovely green, and free from all danger, and is fortunately superseding the Scheele's green both in dresses and in worsted work. I should advise my fair reader, when she selects green as her colour, always to choose the gas-light green, and to wear and to use for worsted work no other green besides, unless it be imperial green.

DIET.

328. *Which is the more wholesome, coffee or tea, where milk does not agree, for a youth's breakfast ?*

Coffee, provided it be made properly, and provided the boy or the girl take a great deal of out-door exercise ; if a youth be much confined within doors, black tea is preferable to coffee. The usual practice of making coffee is to boil it, to get out the strength ! But the fact is, the process of boiling boils the strength away ; it drives off that aromatic, grateful principle, so wholesome to the stomach, and so exhilarating to the spirits ; and, in lieu of which, extracts its dregs and impurities, which are both heavy and difficult of digestion. The coffee ought, if practicable, to be *freshly* ground every morning, in order that you may be quite sure that it be perfectly genuine, and that none of the aroma of the coffee has flown off from long exposure to the atmosphere. If a youth's bowels be inclined to be costive, coffee is preferable to tea for breakfast, as coffee tends to keep the bowels regular. Fresh milk ought always to be added to the coffee in the proportion of half coffee and half new milk. If coffee does not agree, then *black* tea should be substituted, which ought to be taken with plenty of fresh milk in it. Milk may be frequently given in tea, when it otherwise would disagree.

When a youth is delicate, it is an excellent plan to give him, every morning before he leaves his bed, a tumblerful of *new* milk. The draught of milk, of course, is not in any way to interfere with his regular breakfast.

329. *Do you approve of a boy eating meat with his breakfast ?*

This will depend upon the exercise he uses. If he have had a good walk or run before breakfast, or if he intend, after breakfast, to take plenty of athletic out-door exercise, meat, or a rasher or two of bacon, may, with advantage, be eaten ; but not otherwise.

330. *What is the best dinner for a youth ?*

Fresh mutton or beef, a variety of vegetables, and a farinaceous pudding. It is a bad practice to allow him to dine, exclusively, either on a fruit pudding, or on any other pudding, or on pastry. Unless he be ill, he must, if he is to be healthy, strong, and courageous, eat meat every day of his life. "All courageous animals are carnivorous, and greater courage is to be expected in a people, such as the English, whose food is strong and hearty, than in the half-starved commonalty of other countries."—*Sir W. Temple.*

Let him be debarred from rich soups and from high-seasoned dishes, which only disorder the stomach and inflame the blood. It is a mistake to give a boy or a girl broth or soup, in lieu of meat for dinner; the stomach takes such slops in a discontented way, and is not at all satisfied. It may be well, occasionally, to give a youth with his dinner, *in addition to his meat,* either good soup or good broth not highly seasoned, made of good *meat* stock. But after all that can be said on the subject, a plain joint of meat, either roast or boiled, is far superior for health and strength than either soup or broth, let it be ever so good or so well made.

He should be desired to take plenty of time over his dinner, so that he may be able to chew his food well, and thus that it may be reduced to an impalpable mass, and be well mixed with the saliva,—which the action of the jaws will cause to be secreted—before it passes into the stomach. If such were usually the case, the stomach would not have double duty to perform, and a boy would not so frequently lay the foundation of indigestion, &c., which may embitter, and even make miserable, his after-life. Meat, plain pudding, vegetables, bread, and hunger for sauce (which exercise will readily give), is the best, and, indeed, should be, as a rule, the only dinner he should have. A youth ought not to dine later than two o'clock.

331. *Do you consider broths and soups wholesome?*

The stomach can digest solid much more readily than it can liquid food; on which account the dinner, specified above, is far preferable to one either of broth or of soup. Fluids in large quantities too much dilute the gastric juice, and over-distend the stomach, and hence weaken it, and thus produce indigestion : indeed, it might truly be said that the stomach often takes broths and soups in a grumbling way !

332. *Do you approve of a boy drinking beer with his dinner?*

There is no objection to a little good, mild table-beer, but *strong* ale ought never to be allowed. It is, indeed, questionable whether a boy, unless he take unusual exercise, requires anything but water with his meals.

333. *Do you approve of a youth, more especially if he be weakly, having a glass or two of wine after dinner?*

I disapprove of it : his young blood does not require to be inflamed, and his sensitive nerves excited, with wine ; and, if he be delicate, I should be sorry to endeavour to strengthen him by giving him such an inflammable fluid. If he be weakly, he is more predisposed to put on either fever or inflammation of some organ ; and, being thus predisposed, wine would be likely

to excite either the one or the other of them into
action.

"Wine and youth are fire upon fire."—*Fielding.*

A parent ought on no account to allow a boy to touch
spirits, however much diluted; they are, to the young,
still more deadly in their effects than wine.

334. *Have you any objection to a youth drinking tea?*

Not at all, provided it be not *green* tea, that it be not
made strong, and that it have plenty of milk in it.
Green tea is apt to make people nervous, and boys and
girls ought not even to know what it is to be nervous.

335. *Do you object to supper for a youth?*

Meat suppers are highly prejudicial. If he be hungry
(and if he have been much in the open air, he is almost
sure to be), a piece of bread and cheese, or of bread and
butter, with a draught either of new milk or of table
beer, will form the best supper he can have. He ought
not to sup later than eight o'clock.

336. *Do you approve of a boy having anything between meals?*

I do not; let him have four meals a day, and he will
require nothing in the intervals. It is a mistaken notion
that "little and often is best." The stomach requires
rest as much as, or perhaps more than (for it is frequently
sadly over-worked) any other part of the body. I do
not mean that he is to have "*much* and seldom:"
moderation, in everything, is to be observed. Give him
as much as a growing boy requires (*and that is a great
deal*), but do not let him eat gluttonously, as many
indulgent parents encourage their children to do. Intemperance in eating cannot be too strongly condemned.

337. *Have you any objection to a boy having pocket
money?*

It is a bad practice to allow a boy *much* pocket
money; if he be so allowed, he will be loading his
stomach with sweets, fruit, and pastry, and thus his
stomach will become cloyed and disordered, and the keen
appetite, so characteristic of youth, will be blunted, and

ill-health will ensue. "In a public education, boys early learn intemperance, and if the parents and friends would give them less money upon their usual visits, it would be much to their advantage, since it may justly be said that a great part of their disorders arise from surfeit, *'plus occidit gula quam gladius'* (gluttony kills more than the sword)."—*Goldsmith.*

How true is the saying that "many people dig their graves with their teeth." You may depend upon it that more die from stuffing than from starvation! There would be little for doctors to do if there were not so much stuffing and imbibing of strong drinks going on in the world!

AIR AND EXERCISE.

338. *Have you any remarks to make on fresh air and exercise for boys and girls?*

Girls and boys, especially the former, are too much confined within doors. It is imperatively necessary, if you wish them to be strong and healthy, that they should have plenty of fresh air and exercise; remember, I mean fresh air—country air, not the close air of a town. By exercise, I mean the free unrestrained use of their limbs. Girls, in this respect, are unfortunately worse off than boys, although they have similar muscles to develop, similar lungs that require fresh air, and similar nerves to be braced and strengthened. It is not considered lady-like to be natural—all their movements must be measured by rule and compass!

The reason why so many young girls of the present day are so sallow, under-sized, and ill-shaped, is for the want of air and exercise. After a time the want of air and exercise, by causing ill health, makes them slothful and indolent—it is a trouble for them to move from their chairs!

Respiration, digestion, and a proper action of the bowels, imperatively demand fresh air and exercise. Ill health will inevitably ensue if boys and girls are cooped up a great part of the day in a close room. A distinguished writer of the present day says : " The children of the very poor are always out and about. In this respect they are an example to those careful mammas who keep their children, the whole day long, in their chairs, reading, writing, ciphering, drawing, practising music lessons, doing crotchet work, or anything, in fact, except running about, in spite of the sunshine always peeping in and inviting them out of doors ; and who, in the due course of time, are surprised to find their children growing up with· incurable heart, head, lung, or stomach complaints."

339. *What is the best exercise for a youth ?*

Walking or running : provided either of them be not carried to fatigue,—the slightest approach to it should warn a youth to desist from carrying it further. Walking exercise is not sufficiently insisted upon. A boy or a girl, to be in the enjoyment of good health, ought to walk at least ten miles every day. I do not mean ten miles at a stretch, but at different times of the day. Some young ladies think it an awfully long walk if they manage a couple of miles ! How can they, with such exercise, expect to be well? How can their muscles be developed? How can their nerves be braced? How can their spines be strengthened and be straight? How can their blood course merrily through their blood-vessels ? How can their chests expand and be strong ? Why, it is impossible ! Ill health must be the penalty of such indolence, for Nature will not be trifled with ! Walking exercise, then, is the finest exercise that can be taken, and must be taken, and that without stint, if boys and girls are to be strong and well ! The advantage of our climate is, that there is not a day in the whole year that walking exercise cannot be enjoyed. I use the term *enjoyed* advisedly. The roads may, of course, be dirty ; but what of that ! A good thick pair of boots will be the remedy.

Do then, let me entreat you, insist upon your girls and boys taking plenty of exercise ; let them almost live in the open air ! Do not coddle them ; this is a rough world of ours, and they must rough it; they must be knocked about a great deal, and the knocks will do them good. Poor youths who are, as it were, tied to their mother's apron strings, are much to be pitied; they are usually puny and delicate, and effeminate, and utterly deficient of self-reliance.

340. *Do you approve of horse or pony exercise for boys and girls?*

Most certainly I do; but still it ought not to supersede walking. Horse or pony exercise is very beneficial, and cannot be too strongly recommended. One great advantage for those living in towns, which it has over walking, is, that a person may go further into the country, and thus be enabled to breathe a purer and more healthy atmosphere. Again, it is a much more *amusing* exercise than walking, and this, for the young, is a great consideration indeed.

Horse exercise is for both boys and girls a splendid exercise; it improves the figure, it gives grace to the movements, it strengthens the chest, it braces the muscles, and gives to the character energy and courage.

Both boys and girls ought to be early taught to ride. There is nothing that gives more pleasure to the young than riding either on a pony or on a horse, and for younger children, even on that despised, although useful animal, a donkey. Exercise, taken with pleasure, is doubly beneficial.

If girls were to ride more on horseback than they now do, we should hear less of crooked spines and of round shoulders, of chlorosis and of hysteria, and of other numerous diseases of that class, owing, generally, to debility and to mismanagement. Those ladies who "affect the saddle" are usually much healthier, stronger, and straighter than those who either never or but seldom ride on horseback.

Riding on horseback is both an exercise and an amusement, and is peculiarly suitable for the fair sex, more especially as their modes of exercise are somewhat limited, ladies being excluded from following many games, such as cricket, and foot-ball, both of which are practised, with such zest and benefit, by the rougher sex.

341. *Do you approve of carriage exercise?*

There is no muscular exertion in carriage exercise ; its principal advantage is, that it enables a person to have a change of air, which may be purer than the one he is in the habit of breathing. But, whether it be so or not, change of air frequently does good, even if the air be not so pure. Carriage exercise, therefore, does only partial good, and ought never to supersede either walking or horse exercise.

342. *What is the best time of the day, for the taking of exercise?*

In the summer time, early in the morning and before breakfast, as "cool morning air exhilarates young blood like wine." If a boy cannot take exercise upon an empty stomach, let him have a slice of bread and a draught of milk. When he returns home he will be able to do justice to his breakfast. In fine weather he cannot take too much exercise, provided it be not carried to fatigue.

343. *What is the best time for him to keep quiet?*

He ought not to take exercise immediately after—say for half an hour after—a hearty meal, or it will be likely to interfere with his digestion.

AMUSEMENTS.

344. *What amusements do you recommend for a boy as being most beneficial to health?*

Manly games—such as rowing, skating, cricket, quoits, foot-ball, rackets, single-stick, bandy, bowls, skittles, and all gymnastic exercises. Such games bring the muscles into proper action, and thus cause them to be fully developed. They expand and strengthen the chest ; they cause a due circulation of the blood, making it to bound merrily through the blood-vessels, and thus to diffuse health and happiness in its course. Another excellent amusement for boys, is the brandishing of clubs. They ought to be made in the form of a constable's staff, but should be much larger and heavier. The manner of handling them is so graphically described by Addison that I cannot do better than transcribe it :—" When I was some years younger than I am at present, I used to employ myself in a more laborious diversion, which I learned from a Latin treatise of exercises that is written with great erudition ; it is there called the σκιομαχια, or the fighting with a man's own shadow, and consists in the brandishing of two short sticks grasped in each hand, and loaded with plugs of lead at either end. This opens the chest, exercises the limbs, and gives a man all the pleasure of boxing without the blows. I could wish that several learned men would lay out that time which they employ in controversies and disputes about nothing, in this method of fighting with their own shadows. It might conduce very much to evaporate the spleen which makes them uneasy to the public as well as to themselves."

Another capital, healthful game is single-stick, which makes a boy " to gain an upright and elastic carriage, and to learn the use of his limbs."—*H. Kingsley.* Single-stick may be taught by any drill-sergeant in the neighbourhood. Do everything to make a boy strong. Remember, " the glory of young men is their strength."

If games were more patronised in youth, so many miserable, nervous, useless creatures would not abound. Let a boy or girl, then, have plenty of play ; let half of his or her time be spent in play.

There ought to be a gymnasium established in every town of the kingdom. The gymnasium, the cricket ground, and the swimming bath, are among our finest establishments, and should be patronised accordingly.

First of all, by an abundance of exercise and fresh air make your boys and girls strong, and then, in due time, they will be ready and be able to have their minds properly cultivated. Unfortunately, in this enlightened age, we commence at the wrong end—we put the cart before the horse—we begin by cultivating the mind, and we leave the body to be taken care of afterwards ; the results are, broken health, precocious, stunted, crooked, and deformed youths, and premature decay.

One great advantage of gymnastic exercise is, it makes the chest expand, it fills the lungs with air, and by doing so strengthens them amazingly, and wards off many diseases. The lungs are not sufficiently exercised and expanded ; boys and girls, girls especially, do not as a rule half fill their lungs with air ; now air to the lungs is food to the lungs, and portions of the lungs have not half their proper food, and in consequence suffer.

It is very desirable that every boy and girl should, every day of his or her life, and for a quarter of an hour at least each time, go through a regular *breathing exercise*—that is to say, should be made to stand upright, throw back the shoulders, and the while alternately and regularly fully fill and fully empty the lungs of air. If this plan were daily followed, the chest and lungs would be wonderfully invigorated, and the whole body benefited.

345. *Is playing the flute, blowing the bugle, or any other wind instrument, injurious to health ?*

˙ Decidedly so : the lungs and the windpipe are brought into unnatural action by them. If a boy be of a consumptive habit, this will, of course, hold good with tenfold force. If a youth must be musical let him be taught singing, as that, provided the lungs be not diseased, will be beneficial.

346. *What amusements do you recommend for a girl ?*

Archery, skipping, horse exercise, croquet, the handswing, the fly-pole, skating, and dancing, are among the best. Archery expands the chest, throws back the shoulders, thus improving the figure, and develops the muscles. Skipping is exceedingly good exercise for a girl, every part of the body being put into action by it. Horse exercise is splendid for a girl ; it improves the figure amazingly—it is most exhilarating and amusing ; moreover, it gives her courage and makes her self-reliant. Croquet develops and improves the muscles of the arms, beautifies the complexion, strengthens the back, and throws out the chest. Croquet is for girls and women what cricket is for boys and men—a glorious game. Croquet has improved both the health and the happiness of womankind more than any game ever before invented. Croquet, in the bright sunshine, with the winds of heaven blowing about the players, is not like a ball in a stifling hot ball-room, with gas-lights poisoning the air. Croquet is a more sensible amusement than dancing ; it brings the intellect as well as the muscles into play. The man who invented croquet has deserved greater glory, and has done more good to his species, than many philosophers whose names are emblazoned in story. Handswing is a capital exercise for a girl, the whole of the body is thrown into action by it, and the spine, the shoulders, and the shoulder-blades, are especially benefited. The fly-pole, too, is good exercise for the whole of the muscles of the body, especially of the legs and the arms. Skating is for a girl excellent exercise, and is as exhilarating as a glass of champagne, but will do her far more good ! Skating improves the figure, and makes a girl balance and carry herself upright and well ; it is a most becoming exercise for her, and is much in every way to be commended. Moreover, skating gives a girl courage and self-reliance. Dancing, followed as a rational amusement, causes a free circulation of the blood, and provided it does not induce her to sit up late at night, is most beneficial.

347. *If dancing be so beneficial why are balls such fruitful sources of coughs, of colds, and consumptions ?*

On many accounts. They induce young ladies to sit up late at night ; they cause them to dress more lightly than they are accustomed to do ; and thus thinly clad, they leave their homes while the weather is perhaps piercingly cold, to plunge into a suffocating, hot ball-room, made doubly injurious by the immense number of lights, which consume the oxygen intended for the due performance of the healthy functions of the lungs. Their partners, the brilliancy of the scene, and the music, excite their nerves to undue, and thus to unnatural, action, and what is the consequence? Fatigue, weakness, hysterics, and extreme depression follow. They leave the heated ball-room, when the morning has far advanced, to breathe the bitterly cold and frequently damp air of a winter's night, and what is the result? Hundreds die of consumption, who might otherwise have lived. Ought there not, then, to be a distinction between a ball at midnight and a dance in the evening?

348. *But still, would you have a girl brought up to forego the pleasures of a ball ?*

If a parent prefer her so-called pleasures to her health, certainly not ; to such a mother I do not address myself.

349. *Have you any remarks to make on singing, or on reading aloud ?*

Before a mother allows her daughter to take lessons in singing, she should ascertain that there be no actual disease of the lungs, for if there be, it will probably excite it into action ; but if no disease exist, singing or reading aloud is very conducive to health. Public singers are seldom known to die of consumption. Singing expands the chest, improves the pronunciation, enriches the voice for conversation, strengthens the lungs, and wards off many of their diseases.

EDUCATION.

350. *Do you approve of corporal punishments in schools ?*

I do not. I consider it to be decidedly injurious both to body and mind. Is it not painful to witness the pale cheeks and the dejected looks of those boys who are often flogged? If their tempers are mild, their spirits are broken; if their dispositions are at all obstinate, they become hardened and wilful, and are made little better than brutes.* A boy who is often

---

* "I would have given him, Captain Fleming, had he been *my* son," quoth old Pearson the elder, "such a good sound drubbing as he never would have forgotten—never !"

"Pooh ! pooh ! my good sir. Don't tell me. Never saw flog-

flogged loses that noble ingenuousness and fine sensibility so characteristic of youth. He looks upon his school as his prison, and his master as his gaoler, and as he grows up to manhood, hates and despises the man who has flogged him. Corporal punishment is revolting, disgusting, and demoralising to the boy; and is degrad ing to the schoolmaster as a man and as a Christian.

If schoolmasters must flog, let them flog their own sons. If they must ruin the tempers, the dispositions, and the constitution of boys, they have more right to practise upon their own than on other people's children! Oh! that parents would raise—and that without any uncertain sound—their voices against such abominations, and the detestable cane would soon be banished the school-room! "I am confident that no boy," says Addison, "who will not be allured by letters without blows, will never be brought to anything with them. A great or good mind must necessarily be the worse for such indignities; and it is a sad change to lose of its virtue for the improvement of its knowledge. No one has gone through what they call a great school, but must have remembered to have seen children of excellent and ingenuous natures (as have afterwards appeared in their manhood). I say, no man has passed through this way of education but must have seen an ingenuous creature expiring with shame, with pale looks, beseeching sorrow, and silent tears, throw up its honest sighs, and kneel on its tender knees to an inexorable blockhead, to be forgiven the false quantity of a word in making a Latin verse. The child is punished, and the next day he commits a like crime, and so a third, with the same consequence. I would fain ask any reasonable man whether this lad, in the simplicity of his native inno- cence, full of shame, and capable of any impression from that grace of soul, was not fitter for any purpose in this life than after that spark of virtue is extinguished in

gin; in the navy do good. Kept down brutes; never made a man yet."—Dr Norman Macleod in *Good Words*, May 1861.

him, though he is able to write twenty verses in an evening ? "

How often is corporal punishment resorted to at school because the master is in a passion, and he vents his rage upon the poor school-boy's unfortunate back !

Oh ! the mistaken notion that flogging will make a bad-behaved boy a good boy ; it has the contrary effect. " ' I dunno how 'tis, sir,' said an old farm labourer, in reply to a question from his clergyman respecting the bad behaviour of his children, ' I dunno how 'tis ; I beats 'em till they're black and blue, and when they won't kneel down to pray I knocks 'em down, and yet they aint good.' "—*The Birmingham Journal.*

In an excellent article in *Temple Bar* (November 1864) on flogging in the army, the following sensible remarks occur :—" In nearly a quarter of a century's experience with soldiers, the writer has always, and without a single exception, found flogging makes a good man bad, and a bad man worse." With equal truth it may be said that, without a single exception, flogging makes a good boy bad, and a bad boy worse. How many men owe their ferocity to the canings they received when school-boys ! The early floggings hardened and soured them, and blunted their sensibility.

Dr Arnold of Rugby, one of the best schoolmasters that England ever produced, seldom caned a boy—not more than once or twice during the half year ; but when he did cane him, he charged for the use of the cane each time in the bill, in order that the parents might know how many times their son had been punished. At some of our public schools now-a-days, a boy is caned as many times in a morning as the worthy doctor would have caned him during the whole half year ; but then the doctor treated the boys as gentlemen, and trusted much to their honour ; but now many schoolmasters trust much to fear, little to honour, and treat them as brute beasts.

It might be said that the discipline of a school cannot

be maintained unless the boys be frequently caned, that it must be either caning or expulsion. I deny these assertions. Dr Arnold was able to' conduct his school with honour to himself, and with immense benefit to the rising generation, without either frequent canings or expulsions. The humane plan, however, requires at first both trouble and patience; and trouble some school-masters do not like, and patience they do not possess; the use of the cane is quick, sharp, decisive, and at the time effective.

If caning be ever necessary, which it might occasionally be, for the telling of lies for instance, or for gross immorality, let the head master himself be the only one to perform the operation, but let him not be allowed to delegate it to others. A law ought in all public schools to be in force to that effect. High time that something were done to abate such disgraceful practices.

Never should a schoolmaster, or any one else, be allowed, *on any pretence whatever*, to strike a boy upon his head. Boxing of the ears has sometimes caused laceration of the drum of the ear, and consequent partial deafness for life. Boxing of the ears injures the brain, and therefore the intellect.

.It might be said, that I am travelling out of my province in making remarks on corporal chastisement in schools? But, with deference, I reply that I am strictly in the path of duty. My office is to inform you of everything that is detrimental to your children's health and happiness; and corporal punishment is assuredly most injurious both to their health and happiness. It is the bounden duty of every man, and especially of every medical man, to lift up his voice against the abominable, disgusting, and degrading system of flogging, and to warn parents of the danger and the mischief of sending boys to those schools where flogging is, except in rare and flagrant cases, permitted.

351. *Have you any observations to make on the selection of a female boarding-school?*

Home education, where it be practicable, is far preferable to sending a girl to school; as *at* home, her health, her morals, and her household duties, can be attended to much more effectually than *from* home. Moreover, it is a serious injury to a girl, in more ways than one, to separate her from her own brothers: they very much lose their affection for each other, and mutual companionship (so delightful and beneficial between brothers and sisters) is severed.

If home education be not practicable, great care must be taken in making choice of a school. Boarding-school education requires great reformation. Accomplishments, superficial acquirements, and brain-work, are the order of the day; health is very little studied. You ought, in the education of your daughters, to remember that they, in a few years, will be the wives and the mothers of England; and, if they have not health and strength, and a proper knowledge of household duties to sustain their characters, what useless, listless wives and mothers they will make!

Remember, then, the body, and not the mind, ought, in early life, to be principally cultivated and strengthened, and that the growing brain will not bear, with impunity, much book learning. The brain of a school-girl is frequently injured by getting up voluminous questions by rote, that are not of the slightest use or benefit to her, or to any one else. Instead of this ridiculous system, educate a girl to be useful and self-reliant. " From babyhood they are given to understand that helplessness is feminine and beautiful; helpfulness, except in certain received forms of manifestation, unwomanly and ugly. The boys may do a thousand things which are 'not proper for little girls.' "—*A Woman's Thoughts about Women.*

From her twelfth to her seventeenth year, is the most important epoch of a girl's existence, as regards her future health, and consequently, in a great measure, her future happiness; and one, in which, more than at any other period of her life, she requires a plentiful supply of fresh air, exercise, recreation, a variety of innocent amusements, and an abundance of good nourishment—more especially of fresh meat; if therefore you have determined on sending your girl to school, you must ascertain that the pupils have as much plain wholesome nourishing food as they can eat,* that the school be situated in a healthy spot, that it be well-drained, that there be a large play-ground attached to it, that the young people are allowed plenty of exercise in the open air—indeed, that at least one-third of the day is spent there in croquet, skipping, archery, battle-dore and shuttlecock, gardening, walking, running, &c.

Take care that the school-rooms are well-ventilated, that they are not over-crowded, and that the pupils are allowed chairs to sit upon, and not those abominations—forms and stools. If you wish to try the effect of them upon yourselves, sit for a couple of hours without stirring upon a form or upon a stool, and, take my word for it, you will insist that forms and stools be banished for ever from the schoolroom.

Assure yourself that the pupils are compelled to rise early in the morning, and that they retire early to rest; that each young lady has a separate bed ;† and that many are not allowed to sleep in the same room, and that the apartments are large and well-ventilated. In fine, their health and their morals ought to be preferred far above all their accomplishments.

352. *They use, in some schools, straight-backed chairs*

---

* If a girl have an *abundance* of good nourishment, the school-mistress must, of course, be remunerated for the necessary and costly expense; and how can this be done on the paltry sum charged at *cheap* boarding schools? It is utterly impossible! And what are we to expect from poor and insufficient nourishment to a fast-growing girl, and at the time of life, remember, when she requires an *extra* quantity of good sustaining, supporting food? A poor girl, from such treatment, becomes either consumptive or broken down in constitution, and from which she never recovers, but drags out a miserable existence.

† A horse-hair mattress should always be preferred to a feather-bed. It is not only better for the health, but it improves the figure

to make a girl sit upright, and to give strength to her back : do you approve of them ?

Certainly not : the natural and the graceful curve of the back is not the curve of a straight-backed chair. Straight-backed chairs are instruments of torture, and are more likely to make a girl crooked than to make her straight. Sir Astley Cooper ridiculed straight-backed chairs, and well he might. It is always well for a mother to try, for some considerable time, such ridiculous inventions upon herself before she experiments upon her unfortunate daughter. The position is most unnatural. I do not approve of a girl lounging and lolling on a sofa ; but, if she be tired and wants to rest herself, let her, like any other reasonable being, sit upon a comfortable ordinary chair.

If you want her to be straight, let her be made strong ; and if she is to be strong, she must use plenty of exercise and exertion, such as drilling, dancing, skipping, archery, croquet, hand-swinging, horse-exercise, swimming, bowls, &c. This is the plan to make her back straight and her muscles strong. Why should we bring up a girl differently from a boy ? Muscular exercises, gymnastic performances, and health-giving exertion, are unladylike, forsooth !

HOUSEHOLD WORK FOR GIRLS.

353. *Do you recommend household work as a means of health for my daughter ?*

Decidedly : whatever you do, do not make a fine lady of her, or she will become puny and delicate, listless, and miserable. A girl, let her station be what it might, ought, as soon as she be old enough, to make her own bed. There is no better exercise to expand the figure and to beautify the shape than is bed-making. Let her make tidy her own room. Let her use her hands and her arms. Let her, to a great extent, be self-reliant, and let her wait upon herself. There is nothing vulgar in her being useful. Let me ask, Of what use are many girls of the present day ? They are utterly useless. Are they happy ? No, for

the want of employment, they are miserable—I mean
bodily employment, household work. Many girls,
now-a-days, unfortunately, are made to look upon a
pretty face, dress, and accomplishments, as the only
things needed ! And, when they do become women
and wives—if ever they do become women and wives—
what miserable lackadaisical wives, and what senseless,
useless mothers they will make !

CHOICE OF PROFESSION OR TRADE.

354. *What profession or trade would you recommend
a boy of a delicate or of a consumptive habit to follow ?*

If a youth be delicate, it is a common practice among
parents either to put him to some light in-door trade, or,
if they can afford it, to one of the learned professions.
Such a practice is absurd, and fraught with danger.
The close confinement of an in-door trade is highly prejudicial to health. The hard reading requisite to fit a
man to fill, for instance, the sacred office, only increases
delicacy of constitution. The stooping at a desk, in an
attorney's office, is most trying to the chest. The
harass, the anxiety, the disturbed nights, the interrupted
meals, and the intense study necessary to fit a man for
the medical profession, is still more dangerous to health
than either law, divinity, or any in-door trade. "Sir
Walter Scott says of the country surgeon, that he is
worse fed and harder wrought than any one else in the
parish, except it be his horse."—*Brown's Horæ
Subsecivæ.*

A modern writer, speaking of the life of a medical
man, observes, "There is no career which so rapidly
wears away the powers of life, because there is no other
which requires a greater activity of mind and body. He
has to bear the changes of weather, continued fatigue,
irregularity in his meals, and broken rest ; to live in the
midst of miasma and contagion. If in the country, he
has to traverse considerable distances on horseback,
exposed to wind and storm ; to brave all dangers to go
to the relief of suffering humanity. A fearful truth for

medical men has been established by the table of
mortality of Dr Caspar, published in the *British Review*.
Of 1000 members of the medical profession, 600 died
before their sixty-second year ; whilst of persons leading
a quiet life—such as agriculturists or theologians—the
mortality is only 347.　If we take 100 individuals of
each of these classes, 43 theologians, 40 agriculturists,
35 clerks, 32 soldiers, will reach their seventieth year ;
of 100 professors of the healing art, 24 only will reach
that age.　They are the sign-posts to health ; they can
show the road to old age, but rarely tread it themselves."

If a boy, therefore, be of a delicate or of a consumptive habit, an out-door calling should be advised, such
as that of a farmer, of a tanner, or a land-surveyor ; but,
if he be of an inferior station of society, the trade of a
butcher may be recommended.　Tanners and butchers
are seldom known to die of consumption.

I cannot refrain from reprobating the too common
practice among parents of bringing up their boys to the
professions.　The anxieties and the heartaches which
they undergo if they do not succeed (and how can many
of them succeed when there is such a superabundance of
candidates ?) materially injure their health.　" I very
much wonder," says Addison, " at the humour of
parents, who will not rather choose to place their sons
in a way of life where an honest industry cannot but
thrive, than in stations where the greatest probity,
learning, and good sense, may miscarry.　How many
men are country curates, that might have made themselves aldermen of London by a right improvement of a
smaller sum of money than what is usually laid out
upon a learned education ?　A sober, frugal person, of
slender parts and a slow apprehension, might have
thrived in trade, though he starves upon physic ; as a
man would be well enough pleased to buy silks of one
whom he could not venture to feel his pulse.　Vagellius
is careful, studious, and obliging, but withal a little
thick-skulled ; he has not a single client, but might

have had abundance of customers. The misfortune is that parents take a liking to a particular profession, and therefore desire their sons may be of it; whereas, in so great an affair of life, they should consider the genius and abilities of their children more than their own inclinations. It is the great advantage of a trading nation, that there are very few in it so dull and heavy who may not be placed in stations of life which may give them an opportunity of making their fortunes. A well-regulated commerce is not, like law, physic, or divinity, to be overstocked with hands; but, on the contrary, flourishes by multitudes, and gives employment to all its professors. Fleets of merchantmen are so many squadrons of floating shops, that vend our wares and manufactures in all the markets of the world, and find out chapmen under both the tropics."

355. *Then, do you recommend a delicate youth to be brought up either to a profession or to a trade?*

Decidedly : there is nothing so injurious for a delicate boy, or for anyone else, as idleness. Work, in moderation, enlivens the spirits, braces the nerves, and gives tone to the muscles, and thus strengthens the constitution. Of all miserable people, the idle boy, or the idle man, is the most miserable! If you be poor, of course you will bring him up to some calling; but if you be rich, and your boy be delicate (if he be not actually in a consumption), you will, if you are wise, still bring him up to some trade or profession. You will, otherwise, be making a rod for your own as well as for your son's back. Oh, what a blessed thing is work!

SLEEP.

356. *Have you any remarks to make on the sleep of boys and girls?*

Sleeping-rooms, are, generally, the smallest in the house, whereas, for health's sake, they ought to be the largest. If it be impossible to have a *large* bedroom, I should advise a parent to have a dozen or twenty holes (each about the size of a florin) bored with a centre-bit in the upper part of the chamber door, and the same number of holes in the lower part of the door, so as constantly to admit a free current of air from the passages. If this cannot readily be done, then let the bedroom door be left ajar all night, a door chain being on the door to prevent intrusion; and, in the summer time, during the night, let the window-sash, to the extent of about two or three inches, be left open.

If there be a dressing-room next to the bedroom, it will be well to have the dressing-room window, instead of the bedroom window, open at night. The dressing-room door will regulate the quantity of air to be admitted into the bedroom, opening it either little or much, as the weather might be cold or otherwise.

Fresh air during sleep is indispensable to health.—If a bedroom be close, the sleep, instead of being calm and refreshing, is broken and disturbed; and the boy, when he awakes in the morning, feels more fatigued than when he retired to rest.

If sleep is to be refreshing, the air, then, must be pure, and free from carbonic acid gas, which is constantly being evolved from the lungs. If sleep is to be health-giving, the lungs ought to have their proper food—oxygen, and not to be cheated by giving them instead a poison—carbonic acid gas.

It would be well for each boy to have a separate room to himself, and each girl a separate room to herself. If two boys are obliged, from the smallness of the house, to sleep in one room, and if two girls, from the same cause, are compelled to occupy the same chamber, by all means let each one have a *separate* bed to himself and to herself, as it is so much more healthy, and expedient for both boy and girl to sleep alone.

The roof of the bed should be left open—that is to say, the top of the bedstead ought not to be covered with bed furniture, but should be open to the ceiling, in order to encourage a free ventilation of air. A bed-curtain may be allowed on the side of the bed where there are windy currents of air; otherwise bed-curtains and valances ought on no account to be allowed. They prevent a free circulation of the air. A youth should sleep on a horse-hair mattress. Such mattresses greatly improve the figure and strengthen the frame. During the day-time, provided it does not rain, the windows must be thrown wide open, and, directly after he has risen from bed, the clothes ought to be thrown entirely back, in order that they may become, before the bed be made, well ventilated and purified by the air :—

"Do you wish to be healthy?—
　Then keep the house sweet ;
　As soon as you're up
　　Shake each blanket and sheet.

Leave the beds to get fresh.
　On the close crowded floor
Let the wind sweep right through—
　　Open window and door.

The bad air will rush out
　As the good air comes in,
Just as goodness is stronger
　And better than sin.

Do this, it's soon done,
　In the fresh morning air,
It will lighten your labour
　And lessen your care.

You are weary—no wonder,
　There's weight and there's **gloom**
Hanging heavily round
　In each over-full room.

Be sure all the trouble
　Is profit and gain,
For there's head-ache and heart-ache,
　And fever and pain

Hovering round, settling down
　In the closeness and heat ;
Let the wind sweep right through
　Till the air's fresh and sweet,

And more cheerful you'll feel
　Through the toil of the day ;
More refreshed you'll awake
　When the night's passed away. *****

---

* *Household Verses on Health and Happiness.* London : Jarrold and Sons. Every mother should read these *Verses.*

Plants and flowers ought not to be allowed to remain in a chamber at night. Experiments have proved that plants and flowers take up, in the day-time, carbonic acid gas (the refuse of respiration), and give off oxygen (a gas so necessary and beneficial to health), but give out, in the night season, a poisonous exhalation.

Early rising cannot be too strongly insisted upon; nothing is more conducive to health and thus to long life. A youth is frequently allowed to spend the early part of the morning in bed, breathing the impure atmosphere of a bedroom, when he should be up and about, inhaling the balmy and health-giving breezes of the morning :—

> " Rise with the lark, and with the lark to bed :
> The breath of night's destructive to the hue
> Of ev'ry flower that blows. Go to the field,
> And ask the humble daisy why it sleeps
> Soon as the sun departs ? Why close the eyes
> Of blossoms infinite long ere the moon
> Her oriental veil puts off ? Think why,
> Nor let the sweetest blossom Nature boasts
> Be thus exposed to night's unkindly damp.
> Well may it droop, and all its freshness lose,
> Compell'd to taste the rank and pois'nous steam
> Of midnight theatre and morning ball.
> Give to repose the solemn hour she claims ;
> And from the forehead of the morning steal
> The sweet occasion. Oh ! there is a charm
> Which morning has, that gives the brow of age
> A smack of youth, and makes the lip of youth
> Shed perfume exquisite. Expect it not
> Ye who till noon upon a down-bed lie,
> Indulging feverish sleep."—*Hurdis.*

If early rising be commenced in childhood it becomes a habit, and will then probably be continued through life. A boy ought on no account to be roused from his sleep ; but, as soon as he be awake in the morning, he should be encouraged to rise. Dozing—that state between sleeping and waking—is injurious ; it enervates both body and mind, and is as detrimental to health as dram drinking ! But if he rise early he must go to bed betimes ; it is a bad practice to keep him up until the family retire to rest. He ought, winter and summer, to seek his pillow by nine o'clock, and should rise as soon. as he awake in the morning.

Let me urge upon a parent the great importance of *not* allowing the chimney of any bedroom, or of any room in the house, to be stopped, as many are in the habit of doing to prevent, as *they* call it, a draught, but to prevent, as *I* should call it, health.

357. *How many hours of sleep ought a boy to have ?*

This, of course, will depend upon the exercise he takes : but, on an average, he should have every night at least eight hours. It is a mistaken notion that a boy does *better* with *little* sleep. Infants, children, and youths require more than those who are further advanced in years ; hence old people can frequently do with little sleep. This may in a measure be accounted for from the quantity of exercise the young take. Another reason may be, the young have neither racking pain, nor hidden sorrow, nor carking care, to keep them awake ; while, on the contrary, the old have frequently, the one, the other, or all :—

> " Care keeps his watch on every old man's eye,
> And where care lodges, sleep will never lie."—*Shakspeare.*

ON THE TEETH AND THE GUMS.

358. *What are the best means of keeping the teeth and the gums in a healthy state ?*

I would recommend the teeth and the gums to be well brushed with warm salt and water, in the proportion of one large tea-spoonful of salt to a tumbler of water. I was induced to try the above plan by the recommendation of an American writer—*Todd.* The salt and water should be used *every night.*

The following is an excellent tooth-powder :—

Take of—Finely-powder Peruvian Bark ;
    „ Prepared Coral ;
    „ Prepared Chalk ;
    „ Myrrh, of each half an ounce
    „ Orris root, a quarter of an ounce :

Mix them well together in a mortar, and preserve the powder in a wide-mouthed stoppered bottle.

The teeth ought to be well brushed with the above tooth-powder every morning.

If the teeth be much decayed, and if, in consequence, the breath be offensive, two ounces of finely-powdered charcoal well mixed with the above ingredients will be found a valuable addition. Some persons clean their teeth every morning with soap; if soap be used it ought to be Castile soap; and if the teeth be not white and clean, Castile soap is an excellent cleanser of the teeth, and may be used in lieu of the tooth powder as before recommended.

There are few persons who brush their teeth properly. I will tell you the right way. First of all procure a tooth brush of the best make, and of rather hard bristles, to enable it to penetrate into all the nooks and corners of the teeth; then, having put a small quantity of warm water into your mouth, letting the principal of it escape into the basin, dip your brush in warm water, and if you are about using Castile soap, rub the brush on a cake of the soap, and then well brush your teeth, first upwards and then downwards, then from side to side—from right to left, and from left to right—then the backs of the teeth, then apply the brush to the tops of the crowns of the teeth both of the upper and of the lower jaw,—so that *every* part of each tooth, including the gums, may in turn be well cleansed and be well brushed. Be not afraid of using the brush; a good brushing and dressing will do the teeth and the gums an immensity of good; it will make the breath sweet, and will preserve the teeth sound and good. After using the brush the mouth must, of course, be well rinsed out with warm water.

The finest set of teeth I ever saw in my life belonged to a middle-aged gentleman; the teeth had neither spot nor blemish, they were like beautiful pearls. He never had toothache in his life, and did not know what toothache meant! He brushed his teeth, every morning, with soap and water, in the manner I have previously recommended. I can only say to you—go and do likewise!

Camphor, ought never to be used as an ingredient of tooth-powder, it makes the teeth brittle. Camphor certainly has the effect of making the teeth, for a time, look very white; but it is an evanescent beauty.

Tartar is apt to accumulate between and around the teeth; it is better in such a case *not* to remove it by scaling instruments, but to adopt the plan recommended by Dr Richardson, namely, to well brush the teeth with pure vinegar and water.

PREVENTION OF DISEASE, ETC

359 *If a boy or a girl show great precocity of intellect, is any organ likely to become affected?*

A greater quantity of arterial blood is sent to the brain of those who are prematurely talented, and hence it becomes more than ordinarily developed. Such advantages are not unmixed with danger; this same arterial blood may exite and feed inflammation, and either convulsions, or water on the brain, or insanity, or, at last, idiocy may follow. How proud a mother is in having a precocious child! How little is she aware that precocity is frequently an indication of disease!

360. *How can danger in such a case be warded off?*

It behoves a parent, if her son be precocious, to restrain him—to send him to a quiet country place, free from the excitement of the town; and when he is sent to school, to give directions to the master that he is not on any account to tax his intellect (for a master is apt, if he have a clever boy, to urge him forward); and to keep him from those institutions where a spirit of rivalry is maintained, and where the brain is thus kept in a state of constant excitement. Medals and prizes are well enough for those who have moderate abilities, but dangerous, indeed, to those who have brilliant ones.

An over-worked precocious brain is apt to cause the death of the owner; and if it does not do so, it in too many instances injures the brain irreparably, and the possessor of such an organ, from being one of the most

T

intellectual of children becomes one of the most common-
place of men.

Let me urge you, if you have a precocious child, to
give, and that before it be too late, the subject in question your best consideration.

361. *Are precocious boys in their general health usually
strong or delicate ?*

Delicate : nature seems to have given a delicate body
to compensate for the advantages of a talented mind.
A precocious youth is predisposed to consumption, more
so than to any other disease. The hard study which he
frequently undergoes excites the disease into action. It
is not desirable, therefore, to have a precocious child.
A writer in " Fraser's Magazine " speaks very much to
the purpose when he says, " Give us intellectual beef
rather than intellectual veal."

362. *What habit of body is most predisposed to
scrofula ?*

He or she who has a moist, cold, fair, delicate and
almost transparent skin, large prominent blue eyes, pro-
tuberant forehead, light-brown or auburn hair, rosy
cheeks, pouting lips, milk-white teeth, long neck, high
shoulders, small, flat, and contracted chest, tumid bowels,
large joints, thin limbs, and flabby muscles, is the person
most predisposed to scrofula. The disease is not entirely
confined to the above ; sometimes she or he who has black
hair, dark eyes and complexion, is subject to it, but yet,
far less frequently than the former. It is a remarkable
fact that the most talented are the most prone to scrofula,
and being thus clever their intellects are too often
cultivated at the expense of their health. In infancy
and childhood, either water on the brain or mesenteric
disease ; in youth, pulmonary consumption is frequently
their doom : they are like shining meteors ; their life is
short, but brilliant.

363. *How may scrofula be warded off ?*

Strict attention to the rules of health is the means to
prevent scrofula. Books, unless as an amusement, ought
to be discarded. The patient must almost live in the.

open air, and his residence should he a healthy country place, where the air is dry and bracing; if it be at a farm-house, in a salubrious neighbourhood, so much the better. In selecting a house for a patient predisposed to scrofula, *good pure water should be an important requisite;* indeed for every one who values his health. Early rising in such a case is most beneficial. Wine, spirits, and all fermented liquors ought to be avoided. Beef-steaks and mutton-chops in abundance, and plenty of milk and of farinaceous food—such as rice, sago, arrowroot, &c., should be his diet.

Scrofula, if the above rules be strictly and perseveringly followed, may be warded off; but there must be no half measures, no trying to serve two masters—to cultivate at the same time the health and the intellect. The brain, until the body becomes strong, must *not* be taxed. "You may prevent scrofula by care, but that some children are originally predisposed to the disease there cannot be the least doubt, and in such cases the education and the habits of youth should be so directed as to ward off a complaint, the effects of which are so frequently fatal."—*Sir Astley Cooper on Scrofula.*

364. *But suppose the disease to be already formed, what must then be done?*

The plan recommended above must still be pursued, not by fits and starts, but steadily and continuously, for it is a complaint that requires a vast deal of patience and great perseverance. Warm and cold sea-bathing in such a case are generally most beneficial. In a patient with confirmed scrofula it will of course be necessary to consult a skilful and experienced doctor.

But do not allow without a second opinion any plan to be adopted that will weaken the system, which is already too much depressed. No, rather build up the body by good nourishing diet (as previously recommended), by cod-liver oil, by a dry bracing atmosphere, such as, either Brighton, or Ramsgate, or Llandudno; or if the lungs be delicate, by a more sheltered coast, such as, either St Leonards or Torquay.

Let no active purging, no mercurials, no violent, desperate remedies be allowed. If the patient cannot be cured *without* them, I am positive that he will not be cured *with* them.

But do not despair; many scrofulous patients are cured by time and by judicious treatment. But if desperate remedies are to be used, the poor patient had better *by far* be left to Nature : " Let me fall now into the hand of the Lord ; for very great are his mercies ; but let me not fall into the hand of man."—*Chronicles.*

365. *Have you any remarks to make on a girl stooping?*

A girl ought never to be allowed to stoop : stooping spoils the figure, weakens the chest, and interferes with the digestion. If she cannot help stooping, you may depend upon it that she is in bad health, and that a medical man ought to be consulted. As soon as her health is improved the dancing-master should be put in requisition, and calisthenic and gymnastic exercises should be resorted to. Horse exercise and swimming in such a case are very beneficial. The girl should live well, on good nourishing diet, and not be too closely confined either to the house or to her lessons. She ought during the night to lie on a horsehair mattress, and during the day, for two or three hours, flat on her back on a reclining board. Stooping, if neglected, is very likely to lead to consumption.

366. *If a boy be round-shouldered and slouching in his gait, what ought to be done?*

Let him be drilled ; there is nothing more likely to benefit him than drilling. You never see a soldier round-shouldered nor slouching in his gait. He walks every inch like a man. Look at the difference in appearance between a country bumpkin and a soldier ! It is the drilling that makes the difference : " Oh, for a drill-sergeant to teach them to stand upright, and to turn out their toes, and to get rid of that slouching, hulking gait, which gives such a look of clumsiness and stupidity ! "*

---

* A. K. H. B., *Fraser's Magazine,* October 1861,

367. *My daughter has grown out of shape, she has grown on one side, her spine is not straight, and her ribs bulge out more on the one side than on the other; what is the cause, and can anything be done to remedy the deformity ?*

The causes of this lateral curvature of the spine, and consequent bulging out of the ribs that you have just now described, arise either from delicacy of constitution, from the want of proper exercise, from too much learning, or from too little play, or from not sufficient or proper nourishment for a rapidly-growing body. I am happy to say that such a case, by judicious treatment, can generally be cured—namely, by gymnastic exercises, such as the hand-swing, the fly-pole, the patent parlour gymnasium, the chest-expander, the skipping rope, the swimming bath ; all sorts of out-door games, such as croquet, archery, &c. ; by plenty of good nourishment, by making her a child of Nature, by letting her almost live in the open air, and by throwing books to the winds. But let me strongly urge you not, unless ordered by an experienced surgeon, to allow any mechanical restraints or appliances to be used. If she be made strong, the muscles themselves will pull both the spine and the ribs into their proper places, more especially if judicious games and exercises (as I have before advised), and other treatment of a strengthening and bracing nature, which a medical man . will indicate to you, be enjoined. Mechanical appliances will, if not judiciously applied, and in a proper case, waste away the muscles, and will thus increase the mischief ; if they cause the ribs to be pushed in in one place, they will bulge them out in another, until, instead of being one, there will be a series of deformities. No, the giving of strength and the judicious exercising of the muscles are, for a lateral curvature of the spine and the consequent bulging out of one side of the ribs, the proper remedies, and, in the majority of cases, are most effectual, and quite sufficient for the purpose.

I think it well to strongly impress upon a mother's

mind the great importance of early treatment. If the above advice be followed, every curvature in the beginning might be cured. Cases of several years' standing might, with judicious treatment, be wonderfully . relieved.

Bear in mind, then, that if the girl is to be made straight, she is first of all to be made strong ; the latter, together with the proper exercises of the muscles, will lead to the former ; and the *earlier* a medical man takes it in hand, the more rapid, the more certain, and the more effectual will be the cure.

An inveterate, long-continued, and neglected case of curvature of the spine and bulging out of the ribs on one side might require mechanical appliances, but such a case can only be decided on by an experienced surgeon, who ought always, *in the first place*, to be consulted.

368. *Is a slight spitting of blood to be looked upon as a dangerous symptom ?*

Spitting of blood is always to be looked 'upon with suspicion ; even when a youth appears, in other respects, to be in good health, it is frequently the forerunner of consumption. It might be said that, by mentioning the fact, I am unnecessarily alarming a parent, but it would be a false kindness if I did not do so :—

"I must be cruel, only to be kind."—*Shakspeare.*

Let me ask, When is consumption to be cured ? Is it at the onset, or is it when it is confirmed ? If a mother had been more generally aware that spitting of blood was frequently the forerunner of consumption, she would, in the management of her offspring, have taken greater precautions ; she would have made everything give way to the preservation of their health ; and, in many instances, she would have been amply repaid by having the lives of her children spared to her. We frequently hear of patients, in *confirmed* consumption, being sent to Mentone, to Madeira, and to other foreign parts. Can anything be more cruel or absurd ? If there be any disease that requires the comforts of home—and truly may an Englishman's dwelling be called *home!*—and good nursing more than another, it is consumption.

369. *What is the death-rate of consumption in England? At what age does consumption most frequently occur? Are girls more liable to it than boys? What are the symptoms of this disease?*   .

It is asserted, on good authority, that there always are in England, 78,000 cases of consumption, and that the yearly death-rate of this fell disease alone is 39,000! Consumption more frequently shows itself between the ages of fourteen and twenty-one : after then, the liability to the disease gradually diminishes, until, at the age of forty-five, it becomes comparatively rare. Boys are more prone to this complaint than girls. Some of the most important symptoms of pulmonary consumption are indicated by the stethoscope; but, as I am addressing a mother, it would, of course, be quite out of place to treat of such signs in Conversations of this kind. The symptoms it might be well for a parent to recognise, in order that she may seek aid early, I will presently describe. It is perfectly hopeless to expect to cure consumption unless advice be sought at the *onset*, as the only effectual good in this disease is to be done *at first.*

It might be well to state that consumption creeps on insidiously. One of the earliest symptoms of this dreadful scourge is a slight, dry, short cough, attended with tickling and irritation at the top of the throat. This cough generally occurs in the morning; but, after some time, comes on at night, and gradually throughout the day and the night. Frequently during the early stage of the disease *a slight spitting of blood occurs.* Now, this is a most dangerous symptom; indeed, I may go so far as to say that, as a rule, it is almost a sure sign that the patient is in the *first* stage of a consumption.

There is usually hoarseness, not constant, but coming on if the patient be tired, or towards the evening; there is also a sense of lassitude and depression, shortness of breath, a feeling of being quickly wearied—more especially on the slightest exertion. The hair of a con-

sumptive person usually falls off, and what little remains
is weak and poor ; the joints of the fingers become en-
larged, or clubbed as it is sometimes called ; the patient
loses flesh, and, after some time, night sweats make their
appearance : then we may know that hectic fever has
commenced.

Hectic begins with chilliness, which is soon followed
by flushings of the face, and by burning heat of the
hands and the feet, especially of the palms and the soles.
This is soon succeeded by perspirations. The patient has
generally, during the day, two decided paroxysms of
hectic fever—the one at noon, which lasts above five
hours ; the other in the evening, which is more severe,
and ends in violent perspirations, which perspirations
continue the whole night through. He may, during the
day, have several attacks of hectic flushes of the face,
especially after eating ; at one moment he complains of
being too hot, and rushes to the cool air; the next
moment he is too cold, and almost scorches himself by
sitting too near the fire. Whenever the circumscribed
hectic flush is on the cheek, it looks as though the cheek
had been painted with vermilion, then is the time when
the palms of the hands are burning hot. Crabbe, in the
following lines, graphically describes the hectic flush :—

> "When his thin cheek assumed a deadly hue,
> And all the rose to one small spot withdrew :
> They call'd it hectic ; 'twas a fiery flush,
> More fix'd and deeper than the maiden blush."

The expectoration at first is merely mucus, but after a
time it assumes a characteristic appearance ; it has
a roundish, flocculent, woolly form, each portion of
phlegm keeping, as it were, distinct ; and if the expectoration be stirred in water, it has a milk-like appearance.
The patient is commonly harassed by frequent bowel complaints, which rob him of what little strength he has left.
The feet and ankles swell. The perspiration, as before
remarked, comes on in the evening, continues all night
—more especially towards morning, and while the patient
is asleep ; during the time he is awake, even at night, he

seldom sweats much.   The thrush generally shows itself
towards the close of the disease, attacking the tongue, the
tonsils, and the soft palate, *and is a sure harbinger of
approaching death.*   Emaciation rapidly sets in.

If we consider the immense engines of destruction at
work—viz., the colliquative (melting) sweats, the violent
bowel complaints, the vital parts that are affected, the
harassing cough, the profuse expectoration, the hectic
fever, the distressing exertion of struggling to breathe—
we cannot be surprised that " consumption had hung out
her red flag of no surrender," and that death soon closes
the scene.   In girls, provided they have been previously
regular, menstruation gradually declines, and then
entirely disappears.

370. *What are the causes of consumption ?*

The *predisposing* causes of consumption are the
tuberculous habit of body, hereditary predisposition,
narrow or contracted chest, deformed spine, delicacy of
constitution, bad and scanty diet, or food containing but
little nourishment, impure air, close in-door confinement
in schools, in shops, and in factories, ill-ventilated
apartments, dissipation, late hours, over-taxing with
book-learning the growing brain, thus producing
debility, want of proper out-door exercises and amusements, tight lacing; indeed, anything and everything,
that either will debilitate the constitution, or will interfere with, or will impede, the proper action of the lungs,
will be the predisposing causes of this fearful and
lamentable disease.

An ill, poor, and insufficient diet is the mother of
many diseases, and especially of consumption : " Whatsoever was the father of a disease, an ill diet was the
mother."

The most common *exciting* causes of consumption are
slighted colds, neglected inflammation of the chest, long
continuance of influenza, sleeping in damp beds, allowing
wet clothes to dry on the body, unhealthy employ-
ments—such as needle-grinding, pearl button making,
&c.

371. *Supposing a youth to have spitting of bloot,
what precautions would you take to prevent it from ending in consumption ?*

Let his health be the first consideration ; throw books
to the winds ; if he be at school, take him away ; if he
be in trade, cancel his indentures ; if he be in the town,
send him to a sheltered healthy spot in the country, or
to the south coast ; as, for instance, either to St Leonards-
on-Sea, to Torquay, or to the Isle of Wight.

I should be particular in his clothing, taking especial
care to keep his chest and feet warm. If he did not
already wear flannel waistcoats, let it be winter or
summer, I should recommend him immediately to do
so : if it be winter, I should advise him also to take to
flannel drawers. The feet must be carefully attended
to ; they ought to be kept both warm and dry, the
slightest dampness of either shoes or stockings should
cause them to be immediately changed. If a boy, he
ought to wear double-breasted waistcoats ; if a girl, high
dresses.

The diet must be nutritious and generous ; he should
be encouraged to eat plentifully of beef and mutton.
There is nothing better for breakfast, where it agree,
than milk ; indeed, it may be frequently made to agree
by previously boiling it. Good home-brewed ale or
sound porter ought, in moderation, to be taken. Wine
and spirits must on no account be allowed. I caution
parents in this particular, as many have an idea that
wine, in such cases, is strengthening, and that *rum* and
milk is a good thing either to cure or to prevent a
cough !

If it be summer, let him be much in the open air,
avoiding the evening and the night air. If it be winter,
he should, unless the weather be mild for the season,
keep within doors. Particular attention ought to be
paid to the point the wind is in, as he should not be
allowed to go out if it is either in the north, in the east,
or in the north-east ; the latter is more especially
dangerous. If it be spring, and the weather be favour-

able, or summer or autumn, change of air, more especially
to the south-coast—to the Isle of Wight, for instance—
would be desirable; indeed, in a case of spitting of
blood, I know of no remedy so likely to ward off that
formidable, and, generally, intractable complaint—consumption—as change of air. The beginning of the
autumn is, of course, the best season for visiting the
coast. It would be advisable, at the commencement of
October, to send him either to Italy, to the south of
France—to Mentone*—or to the mild parts of England—
more especially either to Hastings, or to Torquay, or to
the Isle of Wight—to winter. But remember, if he be
actually in a *confirmed* consumption, I would not on any
account whatever let him leave his home; as then the
comforts of home will far, very far, out-weigh any benefit
of change of air.

372. *Suppose a youth to be much predisposed to a
sore throat, what precautions ought he to take to ward off
future attacks?*

He must use every morning thorough ablution of the
body, beginning cautiously; that is to say, commencing
with the neck one morning, then by degrees, morning
after morning, sponging a larger surface, until the whole
of the body be sponged. The chill at first must be taken
off the water; gradually the temperature ought to be
lowered until the water be quite cold, taking care to rub
the body thoroughly dry with a coarse towel—a Turkish
rubber being the best for the purpose.

He ought to bathe his throat externally every night
and morning with luke-warm salt and water, the temperature of which must be gradually reduced until at length
no warm water be added. He should gargle his throat
either with barm, vinegar, and sage tea,† or with salt and
water—two tea-spoonfuls of table salt dissolved in a
tumbler of water. He ought to harden himself by taking

* See *Winter and Spring on the Shores of the Mediterranean.* By
J. Henry Bennet, M.D., London : Churchill.

† A wine-glassful of barm, a wine-glassful of vinegar, and
the remainder sage tea, to make a half-pint bottle of gargle.

plenty of exercise in the open air. He must, as much as possible, avoid either sitting or standing in a draught ; if he be in one, he should face it. He ought to keep his feet warm and dry. He should take as little aperient medicine as possible, avoiding especially both calomel and blue-pill. As he grows up to manhood he ought to allow his beard to grow, as such would be a natural covering for his throat : I have known great benefit to arise from this simple plan. The fashion is now to wear the beard, not to use the razor at all, and a sensible fashion I consider it to be. The finest respirator in the world is the beard. The beard is not only good for sore throats, but for weak chests. The wearing of the beard is a splendid innovation ; it saves no end of trouble, is very beneficial to health, and is a great improvement " to the human face divine."

373. *Have you any remarks to make on the almost universal habit of boys and of very young men smoking ?*

I am not now called upon to give an opinion of the effects of tobacco smoking on the middle-aged and on the aged. I am addressing a mother as to the desirability of her sons, when boys, being allowed to smoke. I consider tobacco smoking one of the most injurious and deadly habits a boy or young man can indulge in. It contracts the chest and weakens the lungs, thus predisposing to consumption. It impairs the stomach, thus producing indigestion. It debilitates the brain and nervous system, thus inducing epileptic fits and nervous depression. It stunts the growth, and is one cause of the present race of pigmies. It makes the young lazy and disinclined for work. It is one of the greatest curses of the present day. The following cases prove, more than any argument can prove, the dangerous and deplorable effects of a boy smoking. I copy the first case from *Public Opinion.* "The *France* mentions the following fact as a proof of the evil consequences of smoking for boys :—' A pupil in one of the colleges, only twelve years of age, was some time since seized with epileptic fits, which became worse and worse in spite of all the remedies employed. At last it was discovered that the lad had been for two years past secretly indulging in the weed. Effectual means were adopted to prevent his obtaining tobacco, and he soon recovered.'"

The other case occurred about fifteen years ago in my own practice. The patient was a youth of nineteen. He was an inveterate smoker. From being a bright intelligent lad, he was becoming idiotic, and epileptic fits were supervening. I painted to him, in vivid colours, the horrors of his case, and assured him that if he still persisted in his bad practices, he would soon become a drivelling idiot! I at length, after some trouble and contention, prevailed upon him to desist from smoking altogether. He rapidly lost all epileptic symptoms, his face soon resumed its wonted intelligence, and his mind asserted its former power. He remains well to this day, and is now a married man with a family.

374. *What are the best methods to restrain a violent bleeding from the nose?*

Do not, unless it be violent, interfere with a bleeding from the nose. A bleeding from the nose is frequently an effort of Nature to relieve itself, and therefore, unless it be likely to weaken the patient, ought not to be restrained. If it be necessary to restrain the bleeding, press firmly, for a few minutes, the nose between the finger and the thumb; this alone will often stop the bleeding; if it should not, then try what bathing the nose and the forehead and the nape of the neck with water quite cold from the pump, will do. If that does not succeed, try the old-fashioned remedy of putting a cold large door-key down the back. If these plans fail, try the effects either of powdered alum or of powdered matico, used after the fashion of snuff—a pinch or two either of the one or of the other, or of both, should be sniffed up the bleeding nostril. If these should not answer the purpose, although they almost invariably will, apply a large lump of ice to the nape of the neck, and put a small piece of ice into the patient's mouth for him to suck.

If these methods do not succeed, plunge the hand and the fore-arm into cold water, keep them in for a few minutes, then take them out, and either hold, or let be held up, the arms and the hands high above the head : this plan has frequently succeeded when others have failed. Let the room be kept cool, throw open the windows, and do not have many in the room to crowd around the patient.

Doubtless Dr Richardson's local anæsthetic—the ether spray—playing for a few seconds to a minute *on* the nose and *up* the bleeding nostril, would act most beneficially in a severe case of this kind, and would, before resorting to the disagreeable operation of plugging the nose, deserve a trial. I respectfully submit this suggestion to my medical brethren. The ether—rectified ether—used for the spray ought to be perfectly pure, and of the specific gravity of 0·723.

If the above treatment does not soon succeed, send for a medical man, as more active means, such as plugging of the nostrils—*which is not done unless in extreme cases* —might be necessary.

But before plugging of the nose is resorted to, it will be well to try the effects of a cold solution of alum :—

> Take of—Powdered Alum, one drachm ;
>   Water, half a pint :
> To make a Lotion.

A little of the lotion should be put into the palm of the hand and sniffed up the bleeding nostril ; or, if that does not succeed, some of the lotion ought, by means of a syringe, to be syringed up the nose.

375. *In case of a young lady fainting, what had better be done ?*

Lay her flat upon her back, taking care that the head be as low as, or lower than, the body ; throw open the windows, do not crowd around her,* unloosen her dress

---

* Shakspeare knew the great importance of not crowding around a patient who has fainted. He says—

> "So play the foolish throngs with one that swoons ;
> Come all to help him, and so stop the air
> By which he should revive."

as quickly as possible; ascertain if she have been guilty of tight-lacing—for fainting is sometimes produced by that reprehensible practice. Apply smelling salts to her nostrils; if they be not at hand, burn a piece of rag under her nose; dash cold water upon her face; throw open the window; fan her; and do not, as is generally done, crowd round her, and thus prevent a free circulation of air. As soon as she can swallow, give her either a draught of *cold* water or a glass of wine, or a teaspoonful of sal-volatile in a wine-glassful of water.

To prevent fainting for the future.—I would recommend early hours; country air and exercise; the stays, if worn at all, to be worn slack; attention to diet; avoidance of wine, beer, spirits, excitement, and fashionable amusements.

Sometimes the cause of a young lady fainting, is either a disordered stomach, or a constipated state of the bowels. If the fainting have been caused by *disordered stomach*, it may be necessary to stop the supplies, and give the stomach, for a day or two, but little to do; a fast will frequently prevent the necessity of giving medicine. Of course, if the stomach be *much* disordered, it will be desirable to consult a medical man.

If your daughter's fainting have originated from a *costive state of the bowels* (another frequent cause of fainting), I beg to refer you to a subsequent Conversation, in which I will give you a list of remedies for the prevention and the treatment of constipation.

A young lady's fainting occasionally arises from debility—from downright weakness of the constitution; then the best remedies will be, change of air to the coast, good nourishing diet, and the following strengthening mixture :

Take of—Tincture of Perchloride of Iron, two drachms;
.   Tincture of Calumba, six drachms ;
        Distilled Water, seven ounces :
Two table-spoonfuls of this mixture to be taken three times a day.

Or for a change, the following :–

> Take of—Wine of Iron, one ounce and a-half ;
> Distilled Water, six ounces and a-half
> To make a Mixture. Two table-spoonfuls to be taken three
> times a day.

Iron medicines ought always to be taken *after* instead of *before* a meal. The best times of the day for taking either of the above mixtures will be eleven o'clock, four o'clock, and seven o'clock.

376. *You had a great objection to a mother adminis tering calomel either to an infant or to a child, have you the same objection to a boy or a girl taking it when he or she requires an aperient ?*

Equally as great. It is my firm belief that the frequent use, or rather the abuse, of calomel and of other preparations of mercury, is often a source of liver disease and an exciter of scrofula. It is a medicine of great value in some diseases, when given by a *judicious* medical man; but, at the same time, it is a drug of great danger when either given indiscriminately, or when too often prescribed. I will grant that in liver diseases it frequently gives temporary relief; but when a patient has once commenced the regular use of it, he cannot do without it, until, at length, the *functional* ends in *organic* disease of the liver. The use of calomel predisposes to cold, and thus frequently brings on either inflammation or consumption. Family aperient pills ought never to contain, in any form whatever, a particle of mercury.

377. *Will you give me a list of remedies for the prevention and for the cure of constipation ?*

If you find it necessary to give your son or daughter an aperient, the mildest should be selected; for instance, an agreeable and effectual one, is an electuary composed of the following ingredients :—

> Take of —Best Alexandria Senna, powdered, one ounce
> Best figs, two ounces ;
> Best Raisins (stoned), two ounces ;
> All chopped very fine. The size of a nutmeg or two to be
> eaten, either early in the morning or at bedtime.

Or, one or two tea-spoonfuls of Compound Confection

of Senna (lenitive electuary) may occasionally, early in the morning, be taken. Or, for a change, a tea-spoonful of Henry's Magnesia, in half a tumblerful of warm water. If this should not be sufficiently active, a tea-spoonful of Epsom salts should be given with the magnesia. A Seidlitz Powder forms another safe and mild aperient, or one or two Compound Rhubarb Pills may be given at bed-time. The following prescription for a pill, where an aperient is absolutely necessary, is a mild, gentle, and effective one for the purpose :—

Take of—Extract of Socotrine Aloes, eight grains ;
   Compound Extract of Colocynth, forty-eight grains ;
   Hard Soap, twenty-four grains ;
   Treacle, a sufficient quantity :
To make twenty-four Pills. One or two to be taken at bed-time occasionally.

But, after all, the best opening medicines, are—cold ablutions every morning of the whole body ; attention to diet ; variety of food ; bran-bread ; grapes ; stewed prunes ; French plums ; Muscatel raisins ; figs ; fruit both cooked and raw—if it be ripe and sound ; oatmeal porridge ; lentil powder, in the form of Du Barry's Arabica Revalenta ; vegetables of all kinds, especially spinach ; exercise in the open air ; early rising ; daily visiting the water-closet at a certain hour—there is nothing keeps the bowels open so regularly and well as establishing the habit of visiting the water-closet at a certain hour every morning ; and the other rules of health specified in these Conversations. If more attention were paid to these points, poor school-boys and school-girls would not be compelled to swallow such nauseous and disgusting messes as they usually do to their aversion and injury.

Should these plans not succeed (although in the majority of cases, with patience and perseverance, they will) I would advise an enema once or twice a week, either simply of warm water, or of one made of gruel, table-salt, and olive-oil, in the proportion of two table-spoonfuls of salt, two of oil, and a pint of warm gruel, which a boy may administer to himself, or a girl to herself, by means of a proper enema apparatus.

Hydropathy is oftentimes very serviceable in preventing and in curing costiveness; and as it will sometimes prevent the necessity of administering medicine, it is both a boon and a blessing. "Hydropathy also supplies us with various remedies for constipation. From the simple glass of cold water, taken early in the morning, to the various douches and sea-baths, a long list of useful appliances might be made out, among which we may mention the ' wet compresses ' worn for three hours over the abdomen [bowels], with a gutta percha covering."

I have here a word or two to say to a mother who is always physicking her family. It is an unnatural thing to be constantly dosing either a child, or any one else, with medicine. One would suppose that some people were only sent into the world to be physicked ! If more care were paid to the rules of health, very little medicine would be required ! This is a bold assertion; but I am confident that it is a true one. It is a strange admission for a medical man to make, but, nevertheless, my convictions compel me to avow it.

378. *What is the reason girls are so subject to costiveness ?*

The principal reason why girls suffer more from costiveness than boys, is that their habits are more sedentary; as the best opening medicines in the world are an abundance of exercise, of muscular exertion, and of fresh air. Unfortunately, poor girls in this enlightened age must be engaged, sitting all the while, several hours every day at fancy work, the piano, and other accomplishments; they, consequently, have little time for exercise of any kind. The bowels, as a matter of course, become constipated; they are, therefore, dosed with pills, with black draughts, with brimstone and treacle—Oh ! the abomination !—and with medicines of that class, almost *ad infinitum*. What is the consequence ? Opening medicines, by constant repetition, lose their effects, and, therefore, require to be made stronger and still stronger, until at length, the strongest will scarcely act at all, and the poor unfortunate girl, when she becomes a woman, *if she ever does become one,* is spiritless, heavy, dull, and listless, requiring daily doses of physic, until she almost lives on medicine !

All this misery and wretchedness proceed from Nature's laws having been set at defiance, from *artificial* means taking the place of *natural* ones—from a mother adopting as her rule and guide fashion and folly, rather than reason and common sense. When will a mother awake from her folly and stupidity? This is strong language to address to a lady ; but it is not stronger than the subject demands.

Mothers of England ! do, let me entreat you, ponder well upon what I have said. Do rescue your girls from the bondage of fashion and of folly, which is worse than the. bondage of the Egyptian task-masters ; for the Israelites did, in making bricks without straw, work in the open air—" So the people were scattered abroad throughout all the land of Egypt to gather stubble instead of straw ; " but your girls, many of them, at least, have no work, either in the house or in the open air— they have no exercise whatever. They are poor, drawling, dawdling, miserable nonentities, with muscles, for the want of proper exercise, like ribands ; and with faces, for the lack of fresh air, as white as a sheet of paper. What a host of charming girls are yearly sacrificed at the shrine of fashion and of folly.

Another, and a frequent cause of costiveness, is the bad habit of disobeying the call of having the bowels opened. The moment there is the slightest inclination to relieve the bowels, *instantly* it ought to be attended to, or serious results will follow. Let me urge a mother to instil into her daughter's mind the importance of this advice.

379. *Young people are subject to pimples on the face, what is the remedy ?*

These hard red pimples (*acne*—" the grub pimple ") are a common and an obstinate affection of the skin, affecting the forehead, the temples, the nose, the chin, and the cheeks ; occasionally attacking the neck. the

shoulders, the back, and the chest; and as they more frequently affect the young, from the age of 15 to 35, and
are disfiguring, they cause much annoyance. " These
pimples are so well known by most persons as scarcely
to need description; they are conical, red, and hard;
after a while, they become white, and yellow at the
point, then discharge a thick, yellow-coloured matter,
mingled with a whitish substance, and become covered
by a hard brown scab, and lastly, disappear very slowly,
sometimes very imperfectly, and often leaving an ugly
scar behind them. To these symptoms are not unfrequently added considerable pain, and always much un-
sightliness. When these little cones have the black head
of a 'grub' at their point, they constitute the variety
termed *spotted acne.* These latter often remain stationary
for months, without increasing or becoming red; but
when they inflame, they are in nowise different in their
course from the common kind."—*Wilson on Healthy
Skin.* ·

I find, in these cases, great benefit to be derived from
bathing the face, night and morning, with strong salt and
water—a table-spoonful of table-salt to a tea-cupful of
water; by paying attention to the bowels; by living on
plain, wholesome, nourishing food; and by taking a great
deal of out-door exercise. Sea-bathing, in these cases,
is often very beneficial. Grubs and worms have a
mortal antipathy to salt.

380. *What is the cause of a Gum-boil?*

A decayed root of a tooth, which causes inflammation
and abscess of the gum, which abscess breaks, and thus
becomes a gum-boil.

381. *What is the treatment of a Gum-boil?*

Foment the outside of the face with a hot camomile
and poppy head fomentation,* and apply to the gum-

---

* Four poppy heads and four ounces of camomile blows to be
boiled in four pints of water for half an hour, and then to be
strained to make the fomentation

boil, between the cheek and the gum, a small white bread and milk poultice,* which renew frequently.

As soon as the gum-boil has become quiet, *by all means* have the affected tooth extracted, or it might cause disease, and consequently serious injury of the jaw ; and whenever the patient catches cold there will be a renewal of the inflammation, of the abscess, and of the gum-boil, and, as a matter of course, renewed pain, trouble, and annoyance. Moreover, decayed fangs of teeth often cause the breath to be offensive.

· 382. *What is the best remedy for a Corn ?*

The best remedy for a *hard corn* is to remove it. The usual method of cutting, or of paring a corn away, is erroneous. The following is the right way—Cut with a *sharp* pair of pointed scissors around the circumference of the corn. Work gradually round and round and towards the centre. When you have for some considerable distance well loosened the edges, you can either with your fingers or with a pair of forceps generally remove the corn bodily, and that without pain and without the loss of any blood : this plan of treating a corn I can recommend to you as being most effectual.

If the corn be properly and wholly removed it will leave a small cavity or round hole in the centre, where the blood-vessels· and the nerve of the corn—vulgarly called the root—really were, and which, in point of fact, constituted the very existence or the essence of the corn. Moreover, if the corn be entirely removed, you will, without giving yourself the slightest pain, be able to squeeze the part affected between your finger and thumb.

Hard corns on the sole of the foot and on the sides of the foot are best treated by filing—by filing them with a sharp cutting file (flat on one side and convex

---

* Cut a piece of bread, about the size of the little finger—without breaking it into crumb—pour boiling hot milk upon it, cover it over, and let it stand for five minutes, then apply the soaked bread over the gum-boil, letting it rest between the cheek and the gum.

on the other) neither too coarse nor too fine in the cutting. The corn ought, once every day, to be filed, and should daily be continued until you experience a slight pain, which tells you that the end of the corn is approaching. Many cases of *hard corn* that have resisted every other plan of treatment, have been *entirely* cured by means of the file. One great advantage of the file is, it cannot possibly do any harm, and may be used by a timid person—by one who would not readily submit to any cutting instrument being applied to the corn.

The file, if properly used, is an effectual remedy for a *hard* corn on the sole of the foot. I myself have seen the value of it in several cases, particularly in one case, that of an old gentleman of ninety-five, who had had a corn on the sole of his foot for upwards of half a century, and which had resisted numerous, indeed almost innumerable remedies ; at length I recommended the file, and after a few applications entire relief was obtained, and the corn was completely eradicated.

The corns between the toes are called *soft corns*. A *soft corn* is quickly removed by the strong Acetic Acid—Acid. Acetic Fort.—which ought to be applied to the corn every night by means of a camel's hair brush. The toes should be kept asunder for a few minutes, in order that the acid may soak in ; then apply between the toes a small piece of cotton wool.

Galbanum Plaster spread either on wash leather, or on what is better, on an old white kid glove, has been, in one of our medical journals, strongly recommended as a corn-plaster ; it certainly is an admirable one, and when the corn is between the toes is sometimes most comfortable—affording immense relief.

Corns are like the little worries of life—very teazing and troublesome : a good remedy for a corn—which the Galbanum Plaster undoubtedly is—is therefore worth knowing.

Hard corns, then, on the sole and on the side of the foot are best treated by the file ; *hard corns* on the toes by the scissors; and *soft corns* between the toes either by the strong Acetic Acid or by the Galbanum Plaster.

In the generality of cases the plans recommended above, if properly performed, will effect a cure; but if the corn, from pressure or from any other cause, should return, remove it again, and proceed as before directed. If the corn have been caused either by tight or by ill-fitting shoes, the only way to prevent a recurrence is, of course, to have the shoes properly made by a clever shoemaker—by one who thoroughly understands his business, and who will have a pair of lasts made purposely for the feet.*

The German method of making boots and shoes is a capital one for the prevention of corns, as the boots and shoes are made, scientifically to fit a *real* and not an *ideal* foot.

One of the best preventatives of as well as of the best remedies for corns, especially of soft corns between the toes, is washing the feet every morning, as recommended in a previous Conversation,† taking especial care to wash with the thumb, and afterwards to wipe with the towel between each toe.

383. *What are the best remedies to destroy a Wart?*

As long as fashion instead of common sense, is followed in the making of both boots and shoes, men and women will, as a matter of course, suffer from corns.

It has often struck me as singular, when all the professions and trades are so overstocked, that there should be, as there is in every large town, such a want of chiropodists (corn-cutters) —of respectable chiropodists—of men who would charge a *fixed* sum for every visit the patient may make; for instance to every working-man a shilling, and to every gentleman half-a-crown or five shillings for *each* sitting, and not for *each* corn (which latter system is a most unsatisfactory way of doing business). I am quite sure that if such a plan were adopted, every town of any size in the kingdom would employ regularly one chiropodist at least. However we might dislike some few of the American customs, we may copy them with advantage in this particular —namely, in having a regular staff of chiropodists both in civil and in military life.

† Youth—Ablution, page 250.

Pure nitric acid,* carefully applied to the wart by means of a small stick of cedar wood—a camel's hair pencil-holder—every other day, will soon destroy it. Care must be taken that the acid does not touch the healthy skin, or it will act as a caustic to it. The nitric acid should be preserved in a stoppered bottle and must be put out of the reach of children.

Glacial Acetic Acid is another excellent destroyer of warts : it should, by means of a camel's hair brush, be applied to each wart, every night just before going to bed. The warts will, after a few applications, completely disappear.

384. *What is the best remedy for tender feet, for sweaty feet, and for smelling feet ?*

Cold water : bathing the feet in cold water, beginning with tepid water ; but gradually from day to day reducing the warm until the water be quite cold. A large nursery-basin one-third full of water, ought to be placed on the floor, and one foot at a time should be put in the water, washing the while with a sponge the foot, and with the thumb between each toe. Each foot should remain in the water about half a minute. The feet ought, after each washing, to be well dried, taking care to dry with the towel between each toe. The above process must be repeated at least once every day—every morning, and if the annoyance be great, every night as well. A clean pair of stockings ought in these cases to be put on daily, as perfect cleanliness is absolutely necessary both to afford relief and to effect a cure.

If the feet be tender, or if there be either bunions, or corns, the shoes and the boots made according to the German method (which are fashioned according to the actual shape of the foot) should alone be worn.

385. *What are the causes of so many young ladies of the present day being weak, nervous, and unhappy ?*

---

* A very small quantity of Pure Nitric Acid —just a drain at the bottom of a stoppered bottle—is all that is needed, and which may be procured of a chemist.

The principal causes are—ignorance of the laws of health, Nature's laws being set at nought by fashion and by folly, by want of fresh air and exercise, by want of occupation, and by want of self-reliance. Weak, nervous, and unhappy! Well they might be! What have they to make them strong and happy? Have they work to do to brace the muscles? Have they occupation —useful, active occupation—to make them happy? No! they have neither the one nor the other!

386. *What diseases are girls most subject to?*

The diseases peculiar to girls are—Chlorosis—Green-sickness—and Hysterics.

387. *What are the usual causes of Chlorosis?*

Chlorosis is caused by torpor and debility of the whole frame, *especially of the womb.* It is generally produced by scanty or by improper food, by the want of air and of exercise, and by too close application within doors. Here we have the same tale over again—close application within doors, and the want of fresh air and of exercise! When will the eyes of a mother be opened to this important subject?—the most important that can engage her attention!

388. *What is the usual age for Chlorosis to occur and what are the symptoms?*

Chlorosis more frequently attacks girls from fifteen to twenty years of age; although unmarried women, much older, occasionally have it. I say *unmarried,* for, as a rule, it is a complaint of the *single.*

The patient, first of all, complains of being languid, tired, and out of spirits; she is fatigued with the slightest exertion; she has usually palpitation of the heart (so as to make her fancy that she has a disease of that organ, which, in all probability, she has *not*); she has shortness of breath, and a short dry cough; her face is flabby and pale; her complexion gradually assumes a yellowish or greenish hue—hence the name of chlorosis; there is a dark, livid circle around her eyes; her lips lose their colour, and become almost white; her tongue is generally white and pasty, her appetite is bad, and is frequently depraved—the patient often preferring chalk, slate-pencil, cinder, and even dirt, to the daintiest food; indigestion frequently attends chlorosis ; she has usually pains over the short-ribs, on the *left* side ; she suffers greatly from "wind"—is frequently nearly choked by it ; her bowels are generally costive, and the stools are unhealthy; she has pains in her hips, loins, and back ; and her feet and ankles are oftentimes swollen. *The menstrual discharge is either suspended or very partially performed ;* if the latter, it is usually almost colourless. Hysterical fits not unfrequently occur during an attack of chlorosis.

389. *How may Chlorosis be prevented ?*

If health were more and fashion were less studied, chlorosis would not be such a frequent complaint. This disease generally takes its rise from mismanagement—from Nature's laws having been set at defiance. I have heard a silly mother express an opinion that it is not *genteel* for a girl to eat *heartily !* Such language is perfectly absurd and cruel. How often, too, a weak mother declares that a healthy, blooming girl looks like a milk maid ! It would be well if she did ! How true and sad it is, that "a pale, delicate face, and clear eyes, indicative of consumption, are the fashionable *desiderata* at present for complexion."—*Dublin University Magazine.*

A growing girl requires *plenty* of *good* nourishment—as much as her appetite demands ; and if she have it not, she will become either chlorotic, or consumptive, or delicate. Besides, *the greatest beautifier in the world is health ;* therefore, by a mother studying the health of her daughter, she will, at the same time, adorn her body with beauty ! I am sorry to say that too many parents think more of the beauty than of the health of their girls. Sad and lamentable infatuation ! Nathaniel Hawthorne—a distinguished American—gives a graphic description of a delicate young lady. He says—"She is one of those delicate nervous young creatures not uncommon in New England, and whom I suppose to have become what we find them by the gradually refining away of the physical system among young women,

Some philosophers choose to glorify this habit of body by terming it spiritual; but in my opinion, it is rather the effect of unwholesome food, bad air, lack of out-door exercise, and neglect of. bathing, on the part of these damsels and their female progenitors, all resulting in a kind of hereditary dyspepsia."

Nathaniel Hawthorne was right. Such ladies, when he wrote, were not uncommon ; but within the last two or three years, to their great credit be it spoken, " a change has come o'er the spirit of their dreams," and they are wonderfully improved in health ; for, with all reverence be it spoken, " God helps them who help themselves," and they have helped themselves by attending to the rules of health :—" The women of America are growing more and more handsome every year for just this reason. They are growing rounder of chest, fuller of limb, gaining substance and development in every direction. Whatever may be urged to the contrary we believe this to be a demonstrable fact. . . . When the rising generation of American girls once begin to wear thick shoes, to take much exercise in the open air, to skate, to play at croquet, and to affect the saddle, it not only begins to grow more wise but more healthful, and which must follow as the night the day—more beautiful."—*The Round Table.*

If a young girl had plenty of wholesome meat, varied from day to day, either plain roast or boiled, and neither stewed, nor hashed, nor highly seasoned for the stomach ; if she has had an abundance of fresh air for her lungs ; if she had plenty of active exercise, such as skipping, dancing, running, riding, swimming, for her muscles ; if her clothing were warm and loose, and adapted to the season ; if her mind were more occupied with active *useful* occupation, such as household work, than at present, and if she were kept calm and untroubled from the hurly-burly and excitement of fashionable life— chlorosis would almost be an unknown disease. It is a complaint of rare occurrence with country girls, but of great frequency with fine city ladies.

390. *What treatment should you advise?*

The treatment which would prevent should be adopted when the complaint first makes its appearance. If the above means do not quickly remove it, the mother must then apply to a medical man, and he will give medicines *which will soon have the desired effect.* Chlorosis is very amenable to treatment. If the disease be allowed for any length of time to run on, it may produce either organic —incurable—disease of the heart, or consumption or indigestion, or confirmed ill-health.

391. *At what period of life is a lady most prone to Hysterics, and what are the symptoms?*

The time of life when hysterics occur is generally from the age of fifteen to fifty. Hysterics come on by paroxysms—hence they are called hysterical fits. A patient, just before an attack, is low-spirited ; crying without a cause ; she is "nervous," as it is called ; she has flushings of the face ; she is at other times very pale ; she has shortness of breath and occasional palpitations of the heart ; her appetite is usually bad ; she passes quantities of colourless limpid urine, having the appearance of pump water ; she is much troubled with flatulence in her bowels, and, in consequence, she feels bloated and uncomfortable. The "wind" at length rises upwards towards the stomach, and still upwards to the throat, giving her the sensation of a ball stopping her breathing, and producing a feeling of suffocation. The sensation of a ball in the throat (*globus hystericus*) is the commencement of the fit.

She now becomes *partially* insensible, although she seldom loses *complete* consciousness. Her face becomes flushed, her nostrils dilated, her head thrown back, and her stomach and bowels enormously distended with "wind." After a short time she throws her arms and her legs about convulsively, she beats her breast, tears her hair and clothes, laughs boisterously and screams violently ; at other times she makes a peculiar noise ; sometimes she sobs and her face is much distorted. At length she brings up enormous quantities of wind ; after a time she bursts into a violent flood of tears, and then gradually comes to herself.

As soon as the fit is at an end she generally passes enormous quantities of colourless limpid urine. She might, in a short time, fall into another attack similar to the above. When she comes to herself she feels exhausted and tired, and usually complains of a slight headache, and of great soreness of the body and limbs. She seldom remembers what has occurred during the fit. Hysterics are sometimes frightful to witness ; but, in themselves, are not at all dangerous.

Hysterics—an hysterical fit—is sometimes styled hysterical passion : Shakspeare, in one of his plays, calls it *hysterica passio* :—

"Oh, how this, mother, swells up toward my heart!
Hysterica passio !"

Sir Walter Scott graphically describes an attack :— "The hysterical passion that impels tears is a terrible violence—a sort of throttling sensation—then succeeded by a state of dreaming stupidity."

392. *What are the causes of Hysterics?*

Delicate health, chlorosis, improper and not sufficiently nourishing food, grief, anxiety, excitement of the mind, closely confined rooms, want of exercise, indigestion, flatulence and tight-lacing, are the causes which usually produce hysterics. Hysterics are frequently feigned ; indeed, oftener than any other complaint; and even a *genuine* case is usually much aggravated by a patient herself giving way to them.

393. *What do you recommend an hysterical lady to do ?*

To improve her health by proper management; to rise early and to take a walk, that she may breathe pure and wholesome air,—indeed, she ought to live nearly half her time in the open air, exercising herself with walking, skipping, &c. ; to employ her mind with botany, croquet, archery, or with any out-door amusement ; to confine herself to plain, wholesome, nourishing food ; to avoid tight lacing ; to eschew fashionable amusements ; and, above all, not to give way to her feelings, but, if she feel an attack approaching, to rouse herself.

If the fit be upon her, the better plan is, to banish all the *male* sex from the room, and not even to have many women about her, and for those around to loosen her dress ; to lay her in the centre of the room, flat upon the ground, with a pillow under her head ; to remove combs and pins and brooches from her person ; to dash cold water upon her face ; to apply cloths, or a large sponge wetted in cold water, to her head ; to throw open the window, and then to leave her to herself ; or, at all events, to leave her with only one *female* friend or attendant. If such be done, she will soon come round ; but what is the usual practice ? If a girl be in hysterics, the whole house, and perhaps the neighbourhood, is roused ; the room is crowded to suffocation ; fears are openly expressed by those around that she is in a dangerous state ; she hears what they say, and her hysterics are increased ten-fold.

394. *Have you any remarks to make on a patient recovering from a severe illness ?*

There is something charming and delightful in the feelings of a patient recovering from a severe illness : it is like a new birth : it is almost worth the pain and anguish of having been ill to feel quite well again : everything around and about him wears a charming aspect —a roseate hue : the appetite for food returns with pristine vigour ; the viands, be they ever so homely, never tasted before so deliciously sweet ; and a draught of water from the spring has the flavour of ambrosial nectar : the convalescent treads the ground as though he were on the ambient air ; and the earth to him for a while is Paradise : the very act of living is a joy and gladness :—

> " See the wretch that long has tost
>   On the thorny bed of pain
> Again repair his vigour lost,
>   And walk and run again.

The meanest flow'ret of the vale,
The simplest note that swells the gale,
The common air, the 'earth, the skies,
To him are opening Paradise."—*Gray*

CONCLUDING REMARKS.

If this book is to be of use to mothers and to the rising generation, as I humbly hope and trust that it has been, and that it will be still more abundantly, it ought not to be listlessly read, merely as a novel or as any other piece of fiction; but it must be thoughtfully and carefully studied, until its contents, in all its bearings, be completely mastered and understood.

In conclusion: I beg to thank you for the courtesy, confidence, and attention I have received at your hands; and to express a hope that my advice, through God's blessing, may not have been given in vain; but that it may be—one among many—an humble instrument for improving the race of our children—England's priceless treasures! O, that the time may come, and may not be far distant, "That our sons may grow up as the young plants, and that our daughters may be as the polished corners of the temple!"

INDEX.

x